International Conference on Functional Materials and Chemical Engineering (ICFMCE 2017)

IOP Conference Series: Materials Science and Engineering
Volume 323

Dubai, United Arab Emirates
24 – 26 November 2017

ISBN: 978-1-5108-5928-9
ISSN: 1757-8981

Printed from e-media with permission by:

Curran Associates, Inc.
57 Morehouse Lane
Red Hook, NY 12571

Printed by Curran Associates, Inc. (2018)

For permission requests, please contact the Institute of Physics
at the address below.

Institute of Physics
Dirac House, Temple Back
Bristol BS1 6BE UK

Phone: 44 1 17 929 7481
Fax: 44 1 17 920 0979

techtracking@iop.org

Additional copies of this publication are available from:

Curran Associates, Inc.
57 Morehouse Lane
Red Hook, NY 12571 USA
Phone: 845-758-0400
Fax: 845-758-2633
Email: curran@proceedings.com
Web: www.proceedings.com

Table of Contents

Volume 323

International Conference on Functional Materials and Chemical Engineering (ICFMCE 2017)

24–26 November 2017, Dubai, UAE

Accepted papers received: 21 February 2018
Published online: 2 March 2018

Preface

Papers

IOP Conf. Series: Materials Science and Engineering **323** (2018) 011001 doi:10.1088/1757-899X/323/1/011001

PREFACE

The 1st International Conference on Functional Materials and Chemical Engineering (ICFMCE 2017) was held in Dubai, United Arab Emirates, 24-26 November 2017.

The objective was to provide a forum for professional researchers – scientists and engineers – from the fields of functional materials and chemical engineering as well as mechanical design, manufacture and automation. Further to the usual exchange of knowledge, new scientific concepts, and engineering experiences, which are typical of such gatherings, this event provided opportunities to create new business relationships, learn about cutting-edge research and find partners for future projects. The program included keynote speeches, oral and poster presentations, at a highly specialized level.

The basis for discussion was the material delivered by the following keynote speakers, being internationally recognized academics and leading experts:

- Rafiqul Gani (Technical University of Denmark, Denmark), *A multilayered view of chemical and biochemical engineering*;
- Mario Richard Eden (Auburn University, USA), *Next Generation Computer-Aided Tools for Chemical Process and Product Design*;
- John M Woodley (Technical University of Denmark, Denmark), *Engineering Scalable Biocatalytic Processes*;
- Flavio Manenti (Politecnico di Milano, Italy), *AG2S™ technology: the new route for effective CO_2 reuse*;
- Dong Hwa Kim (Hanbat National University, South Korea), *What will be the core Technology in Mechanical Design, Manufacture and Automation and How will it change: The Paradigm of Society, and Study in 4th Wave Era of 21C*;
- Er Meng Joo (Nanyang Technological University, Singapore), *Cognitive Robotics: Recent Developments and Futuristic Trends*;

And there are also many other impressive speeches delivered by the invited speakers: Edwin Zondervan (Germany), Bent Sarup (Denmark), Yousef Almohairi (UAE), Iqbal M. Mujtaba (UK), Johanna Kleinekorte (Germany), Alberto Pettinau and Enrico Maggio (Italy). The substance of the conferences was comprised of 29 papers from a variety of disciplines, reflecting a range of different points of view, and containing the newest research findings of the authors' teams. The majestic splendor of the meeting location – spectacular, ultra-modern Dubai – complemented the dynamic nature of this prestigious event.

This publication presents the materials covered at the above conferences and serves as both a summary and confirmation of the value of continuing such a congress in future years.

Piotr Kulczycki
On behalf of the Conference Committee

IOP Conf. Series: Materials Science and Engineering **323** (2018) 011002 doi:10.1088/1757-899X/323/1/011002

Peer review statement

All papers published in this volume of *IOP Conference Series: Materials Science and Engineering* have been peer reviewed through processes administered by the proceedings Editors. Reviews were conducted by expert referees to the professional and scientific standards expected of a proceedings journal published by IOP Publishing.

IOP Conf. Series: Materials Science and Engineering **323** (2018) 012001 doi:10.1088/1757-899X/323/1/012001

Low-cost fabrication and performance testing of Polydimethylsiloxane (PDMS) micromixers using an improved print-and-Peel (PAP) method

Ma. Victoria Abagon, Neil Daniel Buendia, Corine Jasper Caracas and Kristian July Yap*

Fuels, Energy, and Thermal Systems Laboratory, Department of Chemical Engineering, University of the Philippines, Diliman, Quezon City, Philippines

*Email: kristianjuly.yap@coe.upd.edu.ph

Abstract.The research presents different configurations of microfluidic mixers made from polydimethylsiloxane (PDMS) fabricated using an improved, low-cost print-and-peel (PAP) method. Processes, such as mixing, operated in the micro scale allow decreased equipment size-to-production capacity ratio and decreased energy consumption per unit product. In the study, saturated solutions of blue and yellow food dyes were introduced inside the channels using a LEGO® improvised microsyringe pump. Scanning Electron Microscopy (SEM) was used to determine the average depth of the fabricated micromixers which was found to be around 14 μm. The flows were observed and images were taken using a light microscope. The color intensities of the images were then measured using MATLAB®. From the relationship between color intensity and concentration, the mixing indices were calculated and found to be 0.9435 to 0.9941, which falls within the standard mixing index range (0.8 – 1.0) regardless of the flow rate and the configuration of the micromixer as verified through the two-way ANOVA. From the cost analysis, the cost of the device fabricated in this study is a hundred-fold less than expenses from standard fabrication procedures. Hence, the fabricated device provides an alternative for micromixers produced from expensive and conventional lithographic methods.

1. Introduction

Microfluidics is a branch of science that deals with the design, fabrication, and manipulation of very small amounts of fluid flow particularly in the microscale [1]. Principles of microfluidics were first utilized in manufacturing inkjet printers during the 1950s, and has extended its applications as of today to various fields such as process intensification, diagnostics, and microelectronics [2].

When working with microfluidic systems, the surface area-to-volume ratio is increased due to its very small characteristic length. Because of the more dominant effects of viscous forces than body forces in these systems, diffusion predominantly works over convection in which components can maximize the effects of diffusion. Processes involved in food processing and pharmaceutical industries which need high viscosity and slow velocities are easier to manipulate in the microscale. Also, the gradients are predictable which gives room for more accurate control over the fluids being mixed [3].

Techniques such as lithographic and non-lithographic methods can be used to fabricate microfluidic devices. Non-lithographic methods offer low-cost fabrication, access to a broader range of materials, and simplicity and flexibility in pattern forming as opposed to photolithography, which is a conventional fabrication technique [4, 5, 6, 7, 8]. Silicon chips are used as the primary material in microfluidic devices. Alternatives to silicon such as polydimethylsiloxane (PDMS) can be utilized to reduce the cost of production. PDMS is usually chosen as the substrate because it is transparent over a wide range of optical frequencies, is biocompatible, and is relatively inexpensive [4].

Conventional fabrication techniques for microfluidic devices such as photolithography are not only costly but also labor-intensive. These methods involve the use of specialized equipment and conditions which limit large-scale production. Developing and improving inexpensive methods will allow more devices to be fabricated easier which will make this technology more accessible to scientists and researchers in low resource countries such as the Philippines.

Some recent researches investigate the ability of office printers in producing the master for microfluidic device fabrication. The desired device pattern is etched onto the master using the printer. PDMS is cured over the pattern to allow the transfer of design from the master to the substrate surface. Reference [4] proposed a low-cost microchannel fabrication by improving the print-and-peel method upon using an inkjet printer. For this proposed microchannel fabrication process, microfluidic devices should be fabricated to demonstrate the validity of this process.

The research aims to fabricate low-cost working micromixers of different configurations and to determine the performance of the fabricated working micromixers. Specifically, this study targets to quantify the mixing index of the fabricated working micromixers through colorimetry, to compare the mixing indices of the fabricated micromixers to existing literature values, and to identify the best configuration among the fabricated micromixers according to their mixing index.

This study is limited to three design configurations used by reference [9]. In fabricating the micromixers, the material used is PDMS and the fabrication method used is adapted from reference [4]. The channel width, aspect ratio, and smoothness of the walls are dependent on AutoCAD® 2015 thinnest preset line and Canon iP 2870. The designs of the mixer were selected according to the limitations of the printer. A glass slide is used as base for the micromixers. The surfaces of both the glass slide and PDMS are subjected to oxygen plasma.

To benchmark the fabricated low-cost micromixers, the fluids used by reference [10] are followed, which are saturated solutions of yellow and blue food dye. These fluids are introduced inside the system using an improvised microsyringe pump made from LEGO® Mindstorms® NXT Kit and programmed using LabVIEW®. The mixing index is determined via colorimetry. Images were taken upon achieving steady-state conditions by the optical microscope. MATLAB® R2011a is used for image processing and analysis. The obtained mixing indices were compared to the standard provided by reference [10].

2. Methodology

2.1. Mold fabrication

Using the thinnest preset line of AutoCAD® 2015, three different micromixer designs from reference [9] are replicated as shown in figure 1.

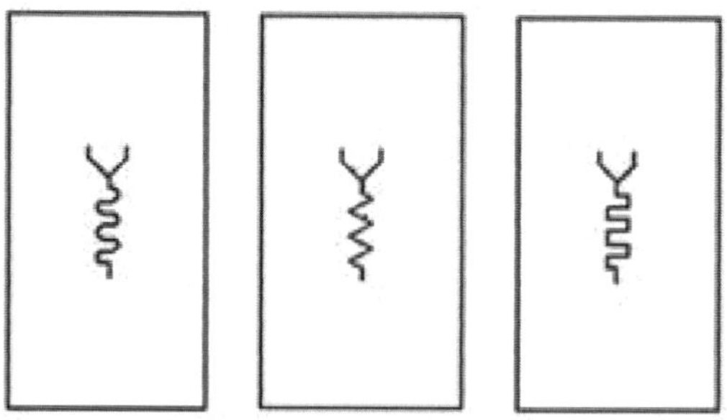

Figure 1. Design configurations (from left to right: curve-wave, zigzag, square-wave).

IOP Conf. Series: Materials Science and Engineering **323** (2018) 012001 doi:10.1088/1757-899X/323/1/012001

The designs are printed on matte sticker paper using a Canon iP 2870 inkjet printer. To increase the depth of the micromixer channels, the designs are repeatedly printed eighty-six (86) times on a single sheet of sticker paper. The operation of the printer is manually modified using a metal wire such that the printing job would not require multiple sheets of paper [4].

The number of overruns, which is the number of times of repeated printing was optimized from reference [4]. The fabricated microchannels are expected to have an aspect ratio of 0.07, which is the average ratio of the depth over the width of the microchannels. The printed molds were left inside a clean bench to dry at atmospheric conditions for at least 12 hours.

2.2. Preparation and curing of PDMS

The PDMS solution was prepared using a 10:1 mass ratio of polymer base-to-curing agent. Considering that the optimum device height is 3 mm based from reference [4], the required mass was obtained by multiplying the density of PDMS (1030 kg/m^3) with the optimum device height of 3 mm and the container area.

For every 6 grams of polymer base, the PDMS solution prepared using a plastic cup and a coffee stirrer is mixed for 2 minutes. Meanwhile, the printed molds are peeled from the sticker paper and placed onto a polystyrene container wrapped with an aluminum foil. The prepared PDMS solution is then evenly poured into the container.

Using a vacuum oven, the PDMS is subjected to a one-hour degassing process to remove trapped air. This is followed by a 90-minute curing process at 100°C under atmospheric conditions. Afterwards, the cured PDMS is cooled to room temperature for 45 minutes.

2.3. Building the micromixer

The cured PDMS, where micromixer channels are now etched, is peeled from the container. Devices are cut accordingly from the whole cured slab of PDMS. At the ends of the channels, holes are created using an improvised puncher where silicon microtubes will be attached.

The adhesion between the surfaces is induced using plasma treatment. Both the channel and the glass slide are washed with ethanol before exposing to oxygen plasma (120 seconds, 320 Pa, 50 W). The exposed surfaces are then made to contact with each other immediately after treatment. Microtubes are then attached to the holes at the ends of the channels using an adhesive.

2.4. Conducting the test runs

Saturated solutions of blue and yellow colored dyes were prepared and loaded to the improvised microsyringe pump constructed using LEGO® Mindstorms®. The mixing is observed using a Euromex Novex optical microscope as the improvised pump introduces the colored dyes into the channels.

3. Results and discussion

3.1. Channel characterization using Scanning Electron Microscopy (SEM)

Samples were cut to get the cross-sectional view of the channels. The exposed cross-section was coated with platinum and subjected to SEM. The width and depth were measured for each sample and the average measurements are presented in table 1.

Table 1. Average channel cross-sectional dimensions.

Micromixer channel configuration	Square-wave	Curve-wave	Zigzag
average channel width, μm	223.12 ± 48.44	186.32 ± 41.14	159.84 ± 33.52
average channel depth, μm	13.36 ± 6.01	13.09 ± 4.11	14.88 ± 3.52

Figure 2 shows some of the SEM images taken in getting the cross-sectional dimensions of the fabricated micromixers.

IOP Conf. Series: Materials Science and Engineering **323** (2018) 012001 doi:10.1088/1757-899X/323/1/012001

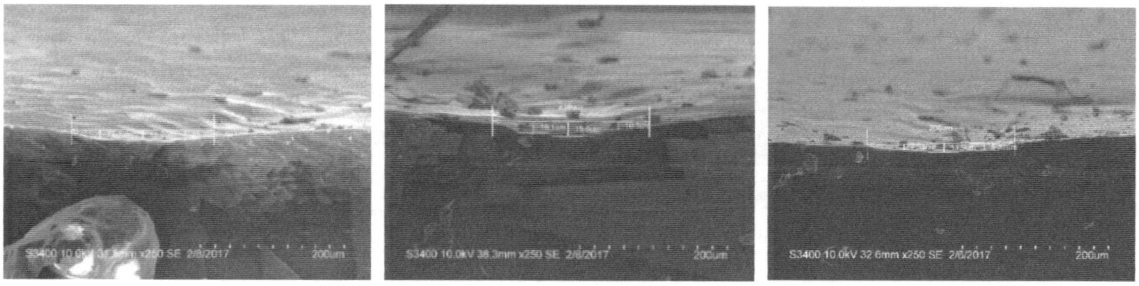

(a) (b) (c)

Figure 2. Cross-section of the channels taken using SEM (a –zigzag, b –curve-wave, c –square-wave).

3.2. Channel characterization using Scanning Electron Microscopy (SEM)

Figure 3 shows the images of the actual flow inside the channels. Analysis is done by subjecting the images of the flow to colorimetry wherein the color intensities are used to describe the mixing behavior inside the channel. The green-colored area between yellow and blue areas is the region where the two fluids mix. The behavior of this mixing region can be related to the extent of diffusive mixing inside the channel.

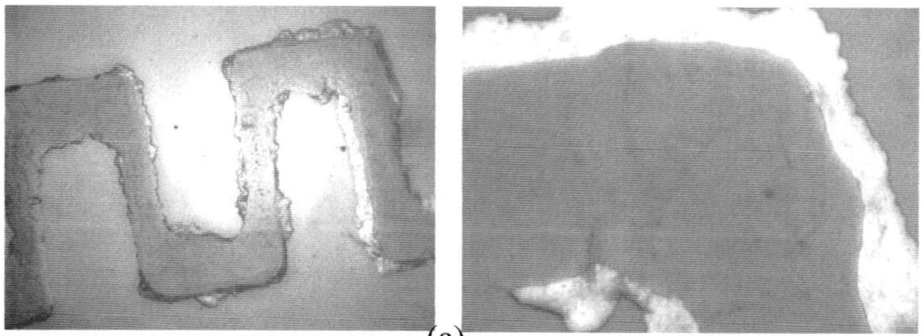

(a)

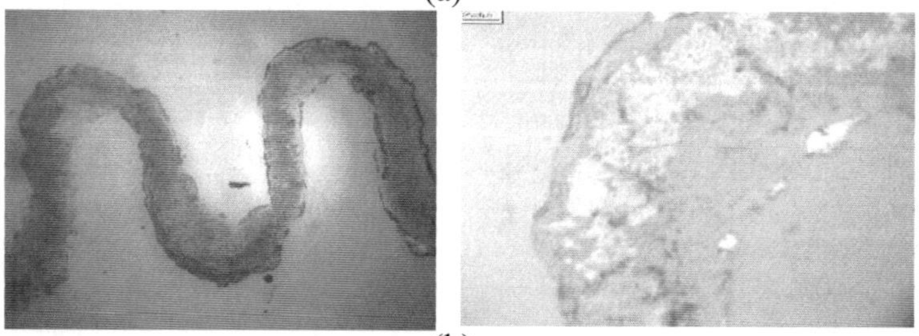

(b)

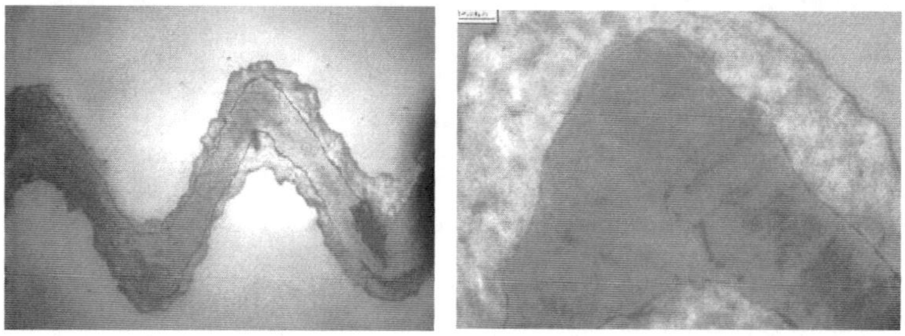

(c)

Figure 3. Images of the actual flow (left at 50x magnification, right at 200x magnification; a –square-wave, b –curve-wave, c –zigzag).

The grayscale intensity value of the green region of each image was determined using MATLAB®. Using this grayscale value, the width of the mixing region was determined by measuring the length of the segment having the same intensity value of the green region. The span of the mixing region was taken at several points along the length of the channel.

Figure 4 shows the graphs of the span of mixing plotted against the projected distance. Generally, the mixing region increases in width along the projected distance inside the channel. The projected distance is the unstretched length of the channels from the first point of contact of the two fluids. This shows that there is an increase in the diffusive mixing as the fluid flows across the channel.

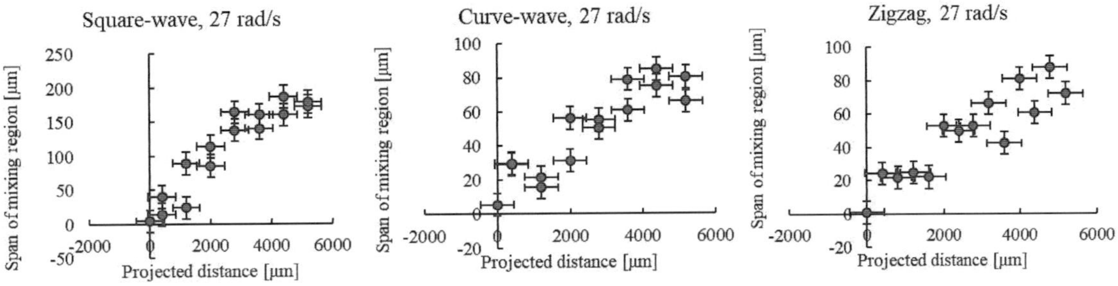

(a) (b) (c)

Figure 4. Span of mixing region plotted against the projected distance (a –square-wave, b –curve-wave, c –zigzag).

The extent of mixing was quantified by finding the mixing index at the end of the channel. Mixing index is a quantification of the extent of mixing ranging from 0, unmixed, to 1, completely mixed. Using the grayscale image of the channel, the color intensity gradient across the width was obtained. The intensity values were then converted to concentration and from the concentration gradient, the mixing index was calculated. Average values are tabulated in table 2.

Table 2. Average mixing indices.

Micromixer channel configuration	Square-wave	Curve-wave	Zigzag
Low-speed (27 rad/s)	0.9745 ± 0.011	0.9893 ± 0.005	0.9785 ± 0.023
Mid-speed (392 rad/s)	0.9774 ± 0.002	0.9755 ± 0.003	0.9782 ± 0.008
High-speed (703 rad/s)	0.9751 ± 0.027	0.9805 ± 0.006	0.9980 ± 0.003

3.3. Hypothesis testing

An F-Test using the two-way ANOVA is conducted to check if a significant relationship exists between the independent variables, (1) mixer configuration and (2) speed at which the fluids are introduced to the mixer, and the dependent variable, which is mixing index.

The following list shows the null hypotheses tested:

- $H_{o,1}$: There is no significant relationship between the mixing index and the speed at which the fluids are introduced.
- $H_{o,2}$: There is no significant relationship between the mixing index and the configuration of the mixers.
- $H_{o,3}$: There is no significant relationship between the mixing index, the speed at which the fluids are induced, and the configuration of the mixers.

Table 3 shows the results of the ANOVA conducted. Upon comparing the F-values calculated and the values from the F-table, it is found that neither of the factors nor its interaction have a significant relationship to the obtained mixing indices.

IOP Conf. Series: Materials Science and Engineering **323** (2018) 012001 doi:10.1088/1757-899X/323/1/012001

Table 3. Calculated F values.

Parameter	Calculated F values	F value from table	Accept/Reject H_0
F_{speed}	0.355	3.555	accept H_0
$F_{configuration}$	0.733	3.555	accept H_0
$F_{interaction}$	0.711	2.983	accept H_0

3.4. Cost analysis and literature comparison
The cost analysis conducted was based on the study of references [4] and [11]. The cost of fabricating a single micromixer using the improved PAP method was compared to the cost of fabricating one wafer of UV LIGA SU-8. The latter was chosen as the representative of photolithography as it is one of the most common techniques used [11]. Equation (1) was used to obtain the cost per unit of device for each mentioned fabrication method:

$$C_M = \frac{C_F}{N} + C_V \qquad (1)$$

where C_M is the cost per unit of a method, C_F is the fixed cost, N is the number of units, and C_V is the variable cost per unit. C_F includes the capital costs of equipment used in fabrication while C_V includes the consumables and operating expenses of the equipment per unit.

Accounted in the capital cost of this study is the inkjet printer. The variable cost consists of the black pigment-based ink, glossy-type sticker paper, and electricity consumed during the printing of the molds and the plasma treatment. Table 4 shows that the improved PAP method is a hundred-fold cheaper compared to conventional photolithography.

Table 4. Cost comparison.

	Improved PAP	UV Liga SU-8
Capital Cost (Php)	1,895	10,209,063
Variable Cost (Php per channel/wafer)	0.15	125.08

In terms of mixing index, the range of mixing indices from the multiple combinations of three different configurations at low, mid and high speed were found to be from 0.9435 to 0.9941. This range falls within the standard mixing index for micromixers which runs from 0.8 to 1.0 [10]. Additionally, a passive micromixer with zigzag channels was reported to have a mixing index of 0.96 [10]. In comparison, the mixing indices obtained from this study are within ±3.5% of 0.96.

4. Conclusion and recommendations
Working micromixers of three different configurations, namely the square-wave, the curve-wave, and the zigzag, have been successfully fabricated. For each configuration, successful mixing of the fluids was observed within the range of speeds induced by the microsyringe pump. Mixing indices were obtained and were found to fall within 0.9435 to 0.9941.

The zigzag configuration gave the highest mixing index values. However, upon conducting F-Test using ANOVA over the pool of data obtained, it is determined that there is no significant relationship between the mixer configuration, the speed induced by the microsyringe pump, and the mixing index.

Given that there is no significant relationship between the factors mentioned, a mixing index within the standard range could be obtained from the micromixers fabricated using this low-cost method, regardless of the mixer configuration and the speed at which the fluids are induced. Additionally, cost analysis shows that fabrication cost per device is 0.15 Php, which is a hundred-fold less than a microfluidic device produced using conventional lithographic methods.

For future studies, it is recommended to work in a cleaner environment to minimize the impurities to which the channels would be exposed. Other configurations and other fluids could also be tested to confirm the validity of the process for the other possible applications of micromixers.

Acknowledgments
We would like to extend our gratitude to the following institutions for making this research possible:
1. *Maynilad Professorial Chair* through the UP Engineering Research Development Foundation, Inc. (ERDFI), Engineering Research and Development for Technology (ERDT), UP Career Assistance Program for Engineering Students (UP CAPES), and the D&L Industries for the financial assistance.
2. Department of Mechanical Engineering (UP Diliman) for the microsyringe pump.
3. Department of Mining, Metallurgical, and Materials Engineering (UP Diliman) for the Plasma Treatment.
4. Polymer Research Laboratory of the Department of Chemical Engineering (UP Diliman) for the optical microscope.
5. Chemical Engineering Analytical Laboratory (CEAL) of UP Diliman for the Scanning Electron Microscope (SEM).

References

[1] Anthonysamy R A 2007 *Designing, Fabricating and Characterizing a Microfluidic Capacitor* (Universiti Malaysia Perlis) pp 4–11
[2] Yadav S 2010 *Analysis of Value Creation and Value Capture in Microfluidics Market* (Massachusetts Institute of Technology)
[3] Nguyen N T 2012 *Micromixers: Fundamentals, Design, and Fabrication* (Massachusetts: William Andrew)
[4] Sanglay J P, Matuba J S, and Abaya R 2016 *Low-Cost Fabrication of a PDMS Microchannel Using an Improved Print-and-Peel (PAP) Method* (University of the Philippines Diliman)
[5] Bao N, Zhang Q, Xu JJ, and Chen HY 2005 *J. Chromatogr. A* 1089 270-75
[6] Quake S and Scherer A 2000 *Science* 290.5496 1536-40
[7] Saikat C, Sharath M, Srujana M, Narayan K and Kumar PP 2015 *Modelling and analysis of microfluidic micromixer for Lab-on-a-Chip (LoC) application* (IEEE)
[8] Vullev V, Wan J, Heinrich V, Landsman P, Bower P, Xia B, Millare B and Jones G 2006 *J. Am. Chem. Soc.* 128 (50) 16062-72
[9] Kuo, J N and Jiang, L R 2014 *Microsys. Technol.* 20 91-99
[10] Viktorov V and Nimafar M 2013 *J. Micromech. Microeng.* 23 1-13
[11] Lawes R 2007 *Microsys. Technol.* 13 85-95

IOP Conf. Series: Materials Science and Engineering **323** (2018) 012002 doi:10.1088/1757-899X/323/1/012002

The 'ideal selectivity' vs 'true selectivity' for permeation of gas mixture in nanoporous membranes

Zhou He and Kean Wang*

Department of Chemical Engineering, Khalifa University of Science & Technology, PO Box 2533, Abu Dhabi, United Arab Emirate.

* Email: kwang@pi.ac.ae.

Abstract. In this study, we proposed and validated a novel and non-destructive experimental technology for measuring the permeation of binary gas mixture in nanoporous membranes. The traditional time lag rig was modified to examine the permeation characteristics of each gas component as well as that of the binary gas mixtures. The difference in boiling points of each species were explored. Binary gas mixtures of CO_2/He were permeated through the nanoporous carbon molecular sieve membrane (CMSM). The results showed that, due to the strong interaction among different molecules and with the porous network of the membrane, the measured perm-selectivity or 'true selectivity' of a binary mixture can significantly deviate from the 'ideal selectivity' calculated form the permeation flux of each pure species, and this deviation is a complicated function of the molecular properties and operation conditions.

1. Introduction

Membrane technology is developing quickly to overtake traditional cryogenic distillation in separation processes, due to its such advantages as: low capital/operation costs, efficiency, easy to scale up and friendly to environment, etc. [1]. Its rapid progress demands new technologies of membrane characterization and evaluation for gas separation applications. Currently, many research publications evaluated the separation performance of a membrane according to the 'ideal selectivity' of a gas pair, which is calculated as the ratio of permeabilities of each pure component measured at similar conditions [2]. Such practice has been found to be inaccurate in many studies, because it represents an oversimplified picture for the complicated diffusion process of gas mixtures in porous media [3]. The traditional time lag [4] technology is popularly used in literature to measure the permeation of pure gas [5]. The diffusion cell technology [6] is capable of measuring the permeation of binary or multicomponent gas mixture, whereas an inert gas is used as the 'sweeping gas' at the permeate side of membrane to carry the permeants to such analytical instruments as gas chromatography (GC) [7], or mass spectrometry (MS) [8]. In addition to the high cost nature of the diffusion cell technology, the presence of the sweeping gas, normally helium, can disturb the system by causing back-diffusion across the membrane. Therefore, it is difficult to measure the 'true' permeation behaviours of each component in membrane separation processes.

To overcome the defects of the existing experimental technologies, this study proposed an improved technology for measuring the 'true' permeation properties of each species in a binary gas mixture. The technique explores the difference in the condensability between two components and is simple in structure, versatile, non-destructive and easy to operate. To validate the feasibility of this technology, inorganic carbon molecular sieve membrane (CMSM) [9] was used as the permeation media while the binary gas mixtures of CO_2/He were prepared at different compositions.

2. Experimental

Figure 1 shows the schematic diagram of the proposed technology, of which traditional time lag rig was modified by adding a cold trap at the U-shape stainless steel tube (1/4 inch in diameter) before the downstream vessel. Liquid N_2 (T = 77K) in the cold trap can condense the heavy component into solid/liquid phase (with constant or negligible vapor pressure) but allows for the light component to reach the downstream vessel for pressure readings. It is seen in table 1 that CO_2 can be easily condensed into solid phase while other gases remained in gas phase. The vapor pressure of solid CO_2 was found to be $\sim 1.2 \times 10^{-6}$ Pa [9] at 77K, which is negligible at the experimental conditions. This means that, by condensing the permeate CO_2 into solid phase, the 'true' permeation behaviour of the other component is revealed. What is more, by adding/removing the cold trap at different time intervals, the 'true' permeation behaviour of the other component as well as the binary mixture can be derived.

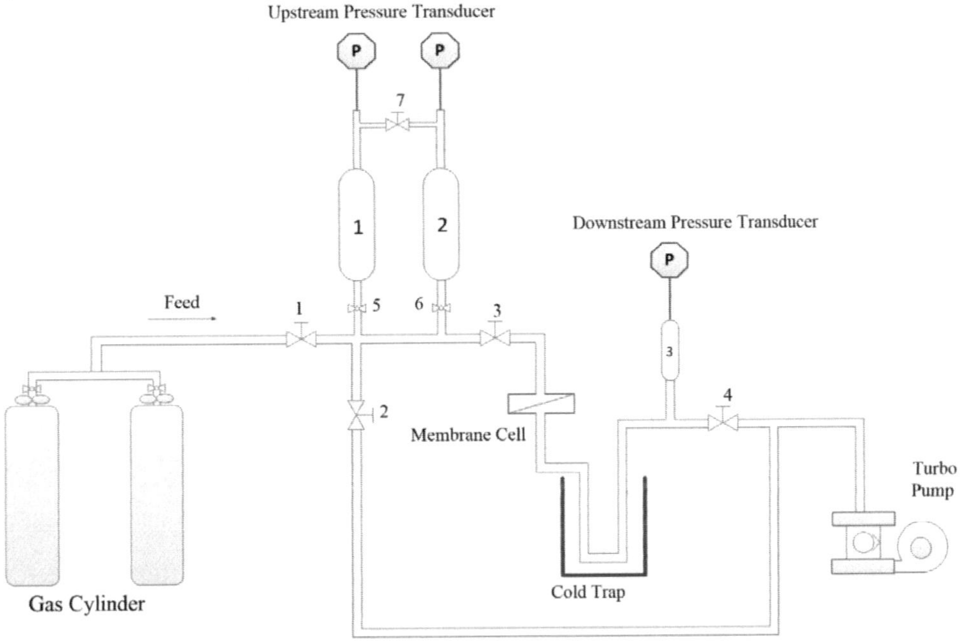

Figure 1. Diagram of improved time lag rig.

Table 1. Properties of CO_2 and He.

Gases	Boiling point (K)	Vapor pressure at 77K	LJ diameters (Å)
He	4.2	----	2.60
CO_2	194.5†	1.2×10^{-6} Pa	3.30

†-CO_2 sublimation point

The permeation experiments were conducted at room temperature on a CMSM derived from Kapton polyimide [10]. To make the gas mixture of required composition (e.g., CO_2:He = 1:1), one specie (He) was firstly charged into vessel 1 (V ~ 500 ml) in figure 1 to designated pressure (e.g., 1bar) and then isolated. The rest of the system was evacuated by vacuuming before the other gas (e.g., CO_2)

IOP Conf. Series: Materials Science and Engineering **323** (2018) 012002 doi:10.1088/1757-899X/323/1/012002

was filled into the vessel 2 to the designated pressure (e.g. 1 bar) and isolated as well. After cleaning the pipeline with vacuum, valve 5, 6 and 7 were opened (while other valves are closed) to mix the two gases in an open loop. A hot air gun was used to slightly heat up one cylinder (up to ~50 °C) to generate forced convection in the open loop to assist the mixing. After two hours of mixing, the system was allowed to cool down to room temperature.

The steady state permeation flux is calculated from the gradient of pressure-time response (dP/dt) of the downstream vessel, as shown in equation (1) while the permeability is calculated in equation (2):

$$J = \frac{V}{SRT} \cdot \frac{dP}{dt} \tag{1}$$

$$Pe = \frac{J}{\Delta P} l \tag{2}$$

where J is the flux (mol/m^2·s), V is the volume of downstream vessel, and S is the cross sectional area of membrane. P_e is the permeability (barrer), ΔP is the pressure difference from upstream to downstream pressure, and l is the thickness of the membrane, respectively.

3. Results and discussions

Figure 2 shows the permeation result of a binary mixture of CO_2-He. The solid line represents the pressure response of downstream vessel under the cold trap while the dotted line shows the response without the cold trap, respectively. We see that the two curved overlap with each other when cold trap was not applied, supporting the reliability of the data.

The detailed procedures are described as follows: At t=0, the cold trap was applied, so that the penetrated CO_2 was condensed into solid phase, while He reached the downstream vessel and gave the pressure response. After reaching the steady state, the 'true' flux of He can be obtained from the gradient of the solid line. Next, at t =3000s, the cold trap was removed and the solid CO_2 would evaporate to gas phase quickly. We see that the dotted line became overlapped with the solid line after the cold trap was removed. The pressure response at this stage would directly correspond to the 'true' permeation of the binary mixture. The steady state CO_2 flux in this binary permeation can be calculated by subtracting the pre-obtained He flux form the total flux. The same process was repeated twice afterwards (by applying and then removing the cold trap), to double check the results and the steady state assumption.

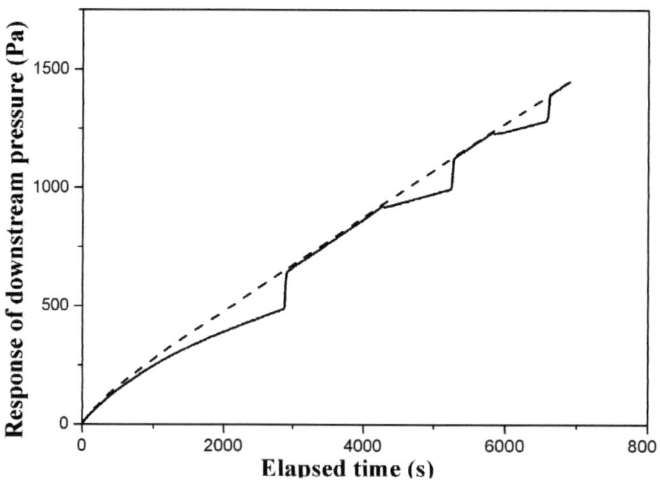

Figure 2. Permeation of gas mixture of CO_2 (1bar) – He (1bar), dash line – without cold trap, solid line – with cold trap.

The permeabilities of He & CO_2 were computed from the experimental results and listed in table 2. The 'ideal' selectivity is defined as the ratio of the permeability of pure species as shown in equation (3):

$$\alpha_{ij} = \frac{Pe(i)}{Pe(j)} \tag{3}$$

The relative permeability of a specie, P_r, is defined as the ratio of its permeability in binary permeation to its permeability in unary permeation [11] at the same partial pressure, as shown in equation (4):

$$P_r = \frac{Pe}{Pe^*} \tag{4}$$

Permeation of pure CO_2 and He were also measured on the rig at similar conditions, respectively, and from which the 'ideal' perm-selectivity was calculated and compared in table 2 with the 'true' selectivity measured with the improved time lag technology.

Table 2. Comparison of the 'True' and 'Ideal' permeation selectivities of CO_2/He.

P_{total} (CO_2: He)	Binary permeation		Unary permeation		α^{True}_{He/CO_2}	α^{Ideal}_{He/CO_2}	P_r (CO_2)	P_r (He)
	Pe (CO_2) (barrer)	Pe (He) (barrer)	Pe* (CO_2) (barrer)	Pe* (He) (barrer)				
2bar (1:1)	31.00	50.70	45.80	144.00	1.64	3.14	0.677	0.352
3bar (2:1)	32.00	30.00	43.00	144.00	0.94	3.35	0.744	0.208
4bar (3:1)	36.00	17.50	40.20	144.00	0.49	3.58	0.896	0.122

It is seen in table 2 that the 'True' selectivity of He/CO_2 can be very different from the 'Ideal' selectivity (up to a factor of 4, as the feed pressure changes) due to the interaction or competitive adsorption-diffusion between two species. This is in agreement with our expectation, as the heavy/slowly diffusing CO_2 molecules [10] can dramatically affect the diffusion flux of weakly adsorbed/light molecules of He. What is more, CO_2 is moderately adsorbed on carbon while He is largely non-adsorbing, which results in a strong surface flux as its concentration increases. This explains our observation that, when the partial pressure of CO_2 increases from 1 bar to 3 bar in the feed, the 'True' perm-selectivity (α^{true}_{He/CO_2}), decreased from ~1.6 to ~0.5, while the 'Ideal' selectivity, (α^{ideal}_{He/CO_2}), remained largely unchanged at ~3.

Next, we look at the relative permeability, P_r, of each component to have an insight into the effect of the interaction between the two species. It is seen in table 2 that: 1) the interaction affect the permeation of both species (as both $P_r < 1$); 2) As the partial pressure of CO_2 in feed increased from 1 bar to 3 bar, P_r of CO_2 was increased by ~30% while that of He was decreased by ~60%. This suggests that this interaction or competitive adsorption-diffusion is a function of chemical potential, molecular mass, and diffusivity, etc. and is more significant for the light species.

All the above discussions suggest that the 'ideal' selectivity [12] can be very misleading when used to evaluate the separation performance of a membrane.

4. Conclusions

The proposed technology is an effective tool for measuring the 'true' permeation properties of binary gas mixtures. The 'ideal' selectivity based on the permeation of pure gas species can significantly deviate from the true values. This deviation is a function of molecular properties (mass, size, diffusivity) and operation conditions (temperature, pressure, concentration, etc.). This research laid solid fundamental for future research in membrane gas separation.

Acknowledgements
We would like to acknowledge the funding from Gas Research Centre (Project code: GRC010, Fund code: 11810/11010) & Abu Dhabi National Oil Company (ADNOC).

References
[1] A. S. Wiryoatmojo, H. Mukhtar, and Z. Man, "Development of polysulfone-carbon molecular sieves mixed matrix membranes for CO_2 removal from natural gas," in 2009 *International Conference on Chemical, Biological and Environmental Engineering* - CBEE 2009, 9-11 Oct. 2009, Singapore, Singapore, 2009, pp. 249-53.
[2] N. Jusoh, Y. F. Yeong, T. L. Chew, K. K. Lau, and A. M. Shariff, "Current Development and Challenges of Mixed Matrix Membranes for CO_2/CH_4 Separation," *Separation and Purification Reviews*, vol. **45**, pp. 321-44, 2016.
[3] S. Li, J. L. Falconer, R. D. Noble, and R. Krishna, "Modeling Permeation of CO_2/CH_4, CO_2/N_2, and N2/CH4 Mixtures Across SAPO-34 Membrane with the Maxwell−Stefan Equations," *Industrial & Engineering Chemistry Research*, vol. **46**, pp. 3904-11, 2007/06/01 2007.
[4] K. Wang, "Diffusion anomaly and blind pore character in carbon molecular sieve membrane," *Chemical Engineering Science*, vol. **62**, pp. 3654-59, Jul 2007.
[5] H. L. Frisch, "The time lag in diffusion," *The Journal of Physical Chemistry*, vol. **61**, pp. 93-95, 1957.
[6] D. M. Ruthven, Principles of adsorption and adsorption processes: *John Wiley & Sons*, 1984.
[7] X. L. Ma, S. Williams, X. T. Wei, J. Kniep, and Y. S. Lin, "Propylene/Propane Mixture Separation Characteristics and Stability of Carbon Molecular Sieve Membranes," *Industrial & Engineering Chemistry Research*, vol. **54**, pp. 9824-31, Oct 2015.
[8] L. J. P. Van Den Broeke and R. Krishna, "Experimental verification of the Maxwell-Stefan theory for micropore diffusion," *Chemical Engineering Science*, vol. **50**, pp. 2507-22, 1995/08/01 1995.
[9] S. Lagorsse, F. D. Magalhães, and A. Mendes, "Carbon molecular sieve membranes: Sorption, kinetic and structural characterization," *Journal of Membrane Science*, vol. **241**, pp. 275-87, 10/1/ 2004.
[10] H. Suda and K. Haraya, "Gas permeation through micropores of carbon molecular sieve membranes derived from Kapton polyimide," *Journal of Physical Chemistry B*, vol. **101**, pp. 3988-94, May 1997.
[11] R. Ash, R. M. Barrer, and C. G. Pope, "Flow of Adsorbable Gases and Vapours in a Microporous Medium. II. Binary Mixtures," *Proceedings of the Royal Society of London. Series A. Mathematical and Physical Sciences,* vol. **271**, p. 19, 1963.
[12] L. M. Robeson, "The upper bound revisited," Journal of Membrane Science, vol. **320**, pp. 390-400, Jul 15 2008.

IOP Conf. Series: Materials Science and Engineering **323** (2018) 012003 doi:10.1088/1757-899X/323/1/012003

Thin film sensor materials for detection of Nitro-Aromatic explosives

Dipali Ramdasi[1] and Rohini Mudhalwadkar[2]

[1]Department of Instrumentation and Control, Cummins College of Engineering for Women, Karvenagar, Pune 411052, INDIA
[2]Department of Instrumentation and Control, College of Engineering Pune, Shivajinagar, Pune 411005, INDIA

*Email: dipali.ramdasi@gmail.com, rpm.instru@coep.ac.in

Abstract. Many countries have experienced terrorist activities and innocent people have suffered. Timely detection of explosives can avoid this situation. This paper targets the detection of Nitrobenzene and Nitrotoluene, which are nitroaromatic compounds possessing explosive properties. As direct sensors for detecting these compounds are not available, Polyaniline based thin film sensors doped with palladium are developed using the spin coating technique. The response of the developed sensors is observed for varying concentrations of explosives. It is observed that zinc oxide based sensor is more sensitive to Nitrotoluene exhibiting a relative change in resistance of 0.78. The tungsten oxide sensor is more sensitive to Nitrobenzene with a relative change in resistance of 0.48. The sensor performance is assessed by measuring the response and recovery time. The cross sensitivity of the sensors is evaluated for ethanol, acetone and methanol which was observed as very low.

1. Introduction

Every country is experiencing terrorist activities and the destruction caused by explosives is well known to all of us. Among hundreds of types of explosives, the use of nitroaromatic explosives is very common. Mixtures with Nitrobenzene (PubChem CID 7416), 4 Nitrotoluene (PubChem CID 7473), are found to be highly explosive of high sensitivity and detonation velocity. Also, mixtures with Nitrobenzene are spark detonable and pose a severe hazard to mankind. [1] The Ion Mobility Spectroscopy, Mass Spectrometry, Infrared Spectroscopy and Raman Spectroscopy are some of the common spectroscopic methods used for explosive detection in-house as described by several researchers [2, 3, 4, 5]. Though these methods are accurate, their use is confined to laboratories. These methods are time consuming and skilled personnel are required to handle these equipment. Though commercially available sensors exhibit a good sensitivity and selectivity, sensors for detecting Nitrobenzene (NB) and Nitrotoluene (NT) are not available. Considering the explosive properties of these compounds, this work aims at developing low cost laboratory developed sensors for detection of Nitrobenzene and Nitrotoluene.

For detection on the field, research on various sensors is being carried out. One of the popular sensors is the metal oxide sensors as suggested by Gardener [6]. Many scientists and engineers have studied metal oxide thin films as electronic materials due to their semiconducting behavior, structural simplicity and low cost [7, 8, 9]. These sensors exhibit change in its electrical properties (typically conductance) when subjected to a gas. This phenomenon occurs due to adsorption and desorption of gas molecules on the gas sensitive surface. The selectivity of the sensor depends on the gas sensitive coating deposited on the sensor surface [10, 11]. The development of simple, low cost reliable sensors for detection of explosives in developing countries like India will accelerate the research towards timely detection of explosives. A critical study was performed to identify the materials to be used for sensor development in the laboratory using spin coating technique [12]. Alexy Tomchenko [13] designed a sensor array consisting of discrete thick-film sensors based on various semiconductor metal oxides (SMO) and fabricated it for ue gas analysis purposes. The selection of the sensitive materials for the array has been accomplished as a result of extensive studies of gas-sensitive properties of SMO. Sofian M. Kanan et. al. and Sergiu T. Shishiyanu [14, 15] presented a study of materials and systems used for detection of various chemical agents and nitro aromatic compounds. This study reveals the use of SnO_2, ZnO, WO_3, CuO, and In_2O_3 as potential materials for nitro aromatic detection. Dan Xie et al [16], developed Polyaniline based sensors for detection of NO_2 and studied the effect of thickness on the sensitivity of film.

2. Materials and Methods

2.1. Development of thin film sensors

After a detailed study of materials used for thin film sensors, sensors were developed with sensing materials of Tin oxide (Research Lab), Zinc oxide (Research Lab), Indium oxide (Research Lab), Polyaniline (Sigma-Aldrich), Lead phthalocyanine (Sigma-Aldrich), Tungsten Oxide (Research Lab). Selectivity is enhanced using palladium metal powder (Ottokemi) with 99.9% purity. The following procedure is adopted to develop the sensors. A micro balance is used to weigh the powders accurately. The weight of the powders depends on the pre-decided ratio of the mixture of powders to achieve a higher selectivity and sensitivity. The weight ratios decided for the sensor is a mixture of selective material powder (6%) and sensitive material powder (94%). The sensitive powder is a mixture of Polyaniline (65%) and metal oxide or phthalocyanine (35%). As the powders need to mix properly and micro or nano granules to form, severe pulverization of the mixture is done. Spin coating machine (Ambala SCM-6VT) is used to form thin film on the glass substrate of size 25mm x 25mm. The pulverized powder is mixed with 2-3 drops of glycerin which acts as a binding agent, before forming the thin film using spin coating machine. 1-2 drops of Dimethyl Sulfoxide are also added, which acts as a chemical agent. Sintering is the process wherein the formed films undergo heat treatment. This hardens the films and makes it quite sturdy. Sintering is carried out at a temperature of 80 degrees Celsius for 10 minutes. High conductivity copper alligator clips are used to generate contacts for the film. The clips are positioned on diagonally opposite sides to reduce contact resistance and thus enhance the response of thin film sensor.

Figure 1 shows a photograph of the developed sensors while figure 2 shows the box with copper alligator clips connected to the sensor. The sensors developed are tested using the Keithly 2110, five and half digit multimeter. The meter can be connected to a computer via USB and the response of each sensor can be stored for further analysis. The sensor response is calculated as the absolute value and expressed in equation (1).

$$\frac{\Delta R}{R} = \frac{R_a - R_g}{R_a} \tag{1}$$

where R_g is the resistance of the sensor when subjected to gas and R_a is the resistance of sensor in air.

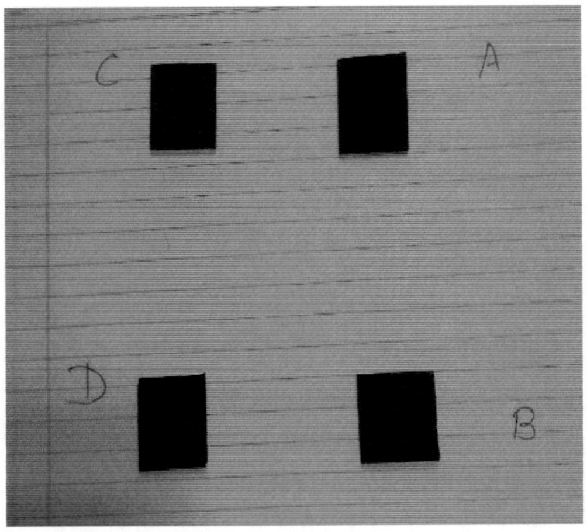

Figure 1. Photograph of some developed sensors.

Figure 2. Photograph of Sensor Box with Copper alligator clips.

2.2. Generation of gas from chemical

To generate the aroma of explosives, Nitrobenzene (LOBA Chemie Pvt. Ltd. Mumbai, India) and Nitrotoluene (Research Lab Fine Chem Industries, Mumbai, India) in chemical form are used. 6 concentrations each of Nitrobenzene and Nitrotoluene, from 1 mL to 30 mL are used as volumes beyond this are dangerous to handle. The nitroaromatic explosive was injected in a glass container of 1 litre capacity and allowed to stand for 25 minutes while covering it. The compound is allowed to mix with normal air and not with a pattern gas. After visually confirming the evaporation, the sensors are subjected to nitro-aroma of the compound through the pump and motor assembly for 60 seconds. Similar process is carried out for generating the volatile organic compounds (VOCs) of Ethanol, Acetone and Methanol (Research Lab Fine Chem Industries, Mumbai, India).

2.3. Experimentation

Experimentation is carried out using the system setup as illustrated in figure 3, while the actual photograph of the setup is shown in figure 4. After acquiring the sensor response for each sensor, the sensor box is purged with air for a pre-decided time interval to avoid interference of traces of earlier gas with the test gas. The sensor response was observed for nitrobenzene and Nitrotoluene using the Keithly multimeter. The cross sensitivity of the tungsten oxide and zinc oxide sensors to ethanol, acetone and methanol is observed using the same setup. The sensor response was also observed for nitro aromatic compounds of Ammonia and Nitric Oxide to study their selectivity towards the intended gas.

3. Results

Ten sensors were developed, out of which Polyaniline based 5 sensors with sensing material of Tin oxide, Zinc oxide, Indium oxide, Lead phthalocyanine and Tungsten Oxide displayed better sensitivity and selectivity towards Nitrobenzene and nitrotoluene. The response of each sensor was observed and recorded using the Keithly 2110, five and a half digit multimeter. An example of the observed response is illustrated in figure 5. The response of the sensors to NB and NT are illustrated in figure 6 and 7. It was observed that the Zinc oxide sensor exhibited maximum sensitivity towards Nitrotoluene and the tungsten oxide sensor showed maximum sensitivity towards Nitrobenzene. The cross sensitivity of the tungsten oxide and zinc oxide sensors is studied for Ethanol, Acetone and Methanol. The experimentation carried out followed the same procedure as that for nitrobenzene and nitrotoluene. It is observed that both the sensors were less sensitive to these VOCs. The response of these sensors is presented in figure 8 and 9. Table 1 shows the sensor performance in terms of response time, recovery time and sensitivity.

IOP Conf. Series: Materials Science and Engineering **323** (2018) 012003 doi:10.1088/1757-899X/323/1/012003

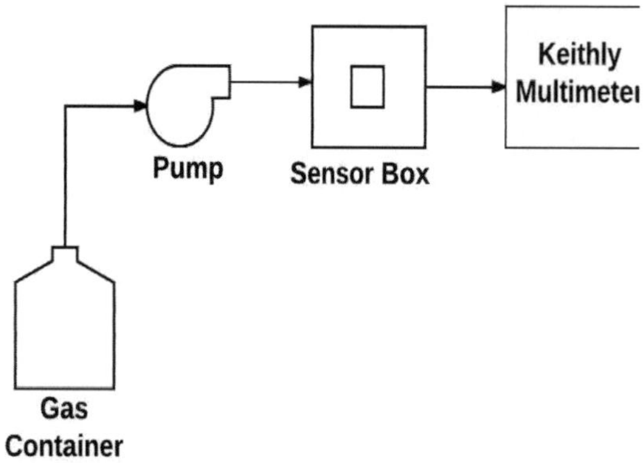

Figure 3. System schematic Diagram.

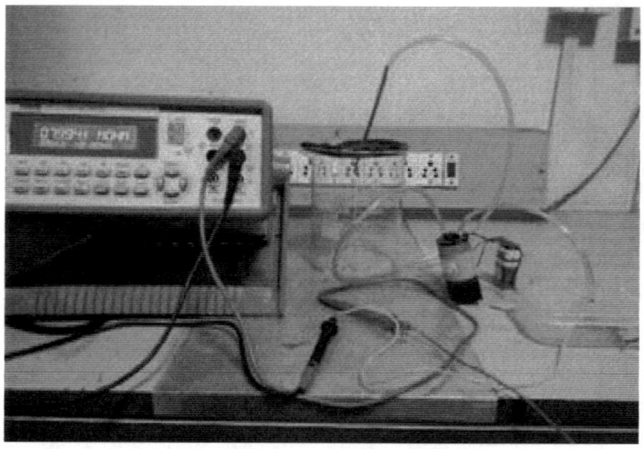

Figure 4. Photograph of the system.

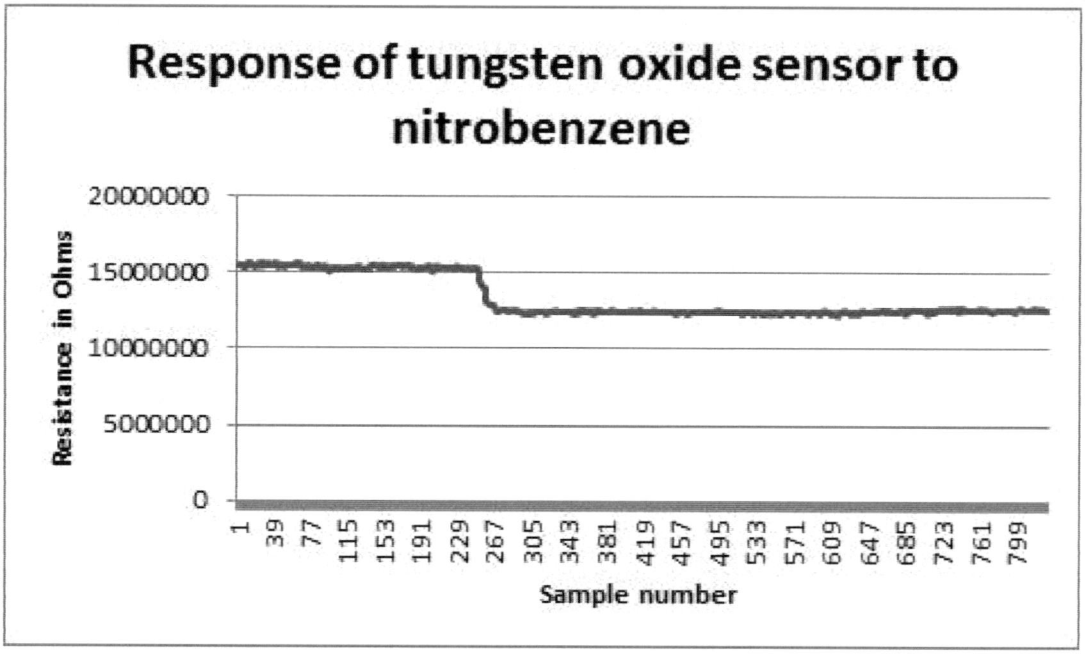

Figure 5. Response of tungsten oxide sensor to Nitrobenzene.

Table 1. Sensor performance.

Sensor Material	Response time (Seconds)	Recovery time (Seconds)	Percentage Sensitivity Nitrobenzene	Percentage Sensitivity Nitrotoluene
Tungsten Oxide	24	360	48	21
Indium Oxide	38	450	13	24
Zinc oxide	52	510	8.6	7.8
Lead Phthalocyanine	12	205	24	34.7
Tin Oxide	43	280	3	38

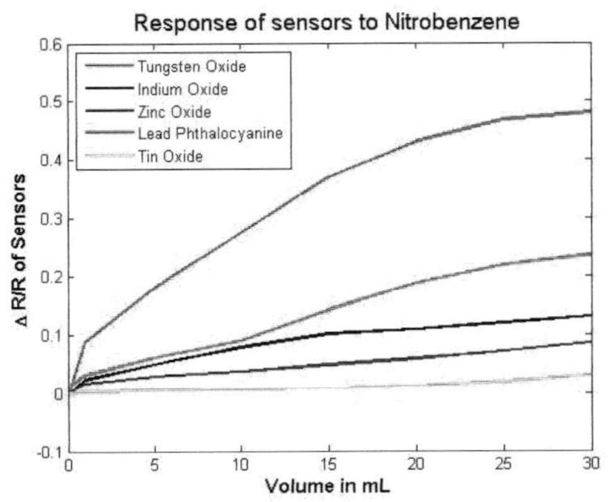

Figure 6. Response of sensors to Nitrobenzene.

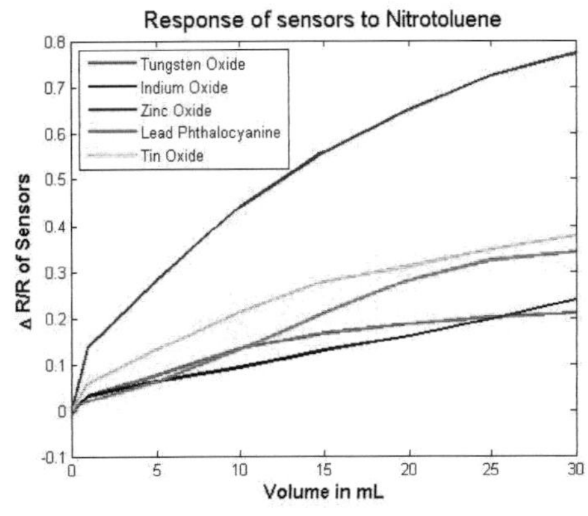

Figure 7. Response of sensors to Nitroluene.

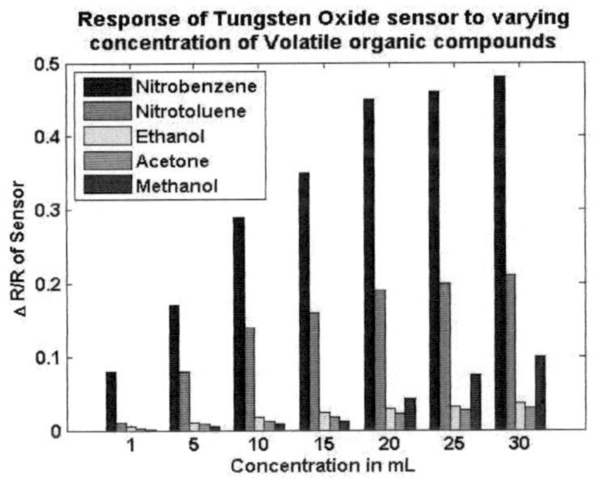

Figure 8. Response of Tungsten Oxide sensor to varying concentrations of Volatile Organic Compounds.

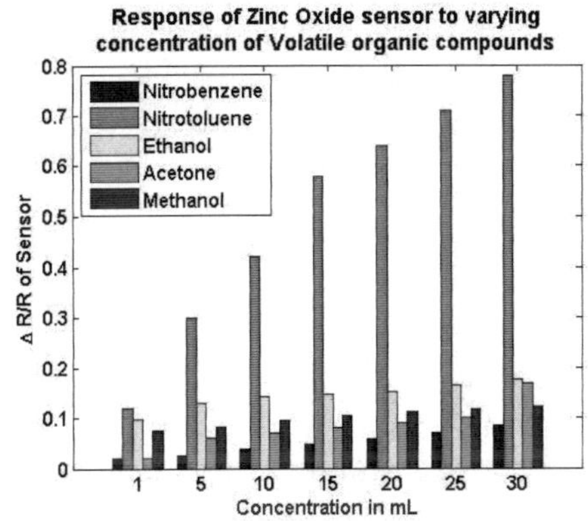

Figure 9. Response of Zinc Oxide sensor to varying concentrations of Volatile Organic Compounds.

4. Discussion and Conclusion

Study reveals the explosive properties of Nitrobenzene and Nitrotoluene and the literature reveals limited availability of sensors to detect these nitroaromatic explosive compounds. Thus thin film sensors are developed in the laboratory using the spin coating technique for detection of these compounds. Five sensors with sensing material of Tin oxide, Zinc oxide, Indium oxide, Lead phthalocyanine and Tungsten Oxide exhibited a measurable sensitivity towards these compounds. It was observed that though the sensors responded to both the compounds, their sensitivity was different. Also, it is observed that zinc oxide based sensor exhibited maximum sensitivity towards Nitrotoluene while tungsten oxide exhibited maximum sensitivity towards Nitrobenzene. Very low cross sensitivity was observed for the tungsten oxide and zinc oxide sensors towards ethanol, acetone and methanol. Extended work can be performed towards the study and improvement of repeatability of these sensors. The development of these sensors will help in the timely detection of explosives. As the method of developing these sensors is simple and of low cost, the sensor response can be acquired for various nitro aromatic compounds. These responses can be included in parametric models developed for explosive detection.

Acknowledgments

This project is funded by Bhabha Atomic Research Centre, India under BRNS scheme, file number 2012/36/55/BRNS/March 2013.

IOP Conf. Series: Materials Science and Engineering **323** (2018) 012003 doi:10.1088/1757-899X/323/1/012003

References

[1] Bretherick L 2016 Bretherick's handbook of reactive chemical hazards (Elsevier)
[2] Ewing R, Atkinson D, Eiceman G and Ewing G 2001 Talanta 54-3 515
[3] Hatab N, Eres G, Hatzinger P and Gu B 2010 J. Raman Spect., 41-10 1131
[4] Primera-Pedrozo O, Soto-Feliciano Y, Pacheco-Londono L and Hernndez-Rivera S 2009 Int. J. Sens. Imag. 10(1-2): 101-13
[5] Sanders N, Kothari S, Huang G, Salazar G and Cooks R 2010 J. Anal. Chem. 82-12 5313
[6] Gardner J and Yinon J 2004 Electronic noses & sensors for the detection of explosives (Kluwer Academic Publishers)
[7] Toko K 2013 Biochemical sensors: mimicking gustatory and olfactory senses (CRC Press)
[8] Vattappalam S, Thomas D, Augustine S and Mathew S 2015 J, Mat. Sci. Eng.: Conf. Series 73-101 2106
[9] Zeggar M, Bourfaa F, Adjimi A, Aida M and Attaf N 2016 J, Mat. Sci. Eng.: Conf. Series 108-1 01 2004
[10] Ramgir N, Mulla I and Vijayamohanan K 2005 Sens.and Act. B 107-2 708
[11] Bai H. and Shi G. 2007 Sensors 7-3 267
[12] Nicolas-Debarnot D. and Poncin-Epaillard F 2003 Anal. Chim. Acta 475-1 1
[13] Tomchenko A., Harmer G, Marquis B and Allen J 2003 Sens.and Act. B 93-1 126
[14] Kanan S, El-Kadri O, Abu-Yousef I and Kanan M 2009 Sensors 9-108 158
[15] Shishiyanu S, Shishiyanu T and Lupan O 2005 Sens.and Act. B 107-1 379
[16] Xie D, Jiang Y, Pan W, Li D, Wu Z and Li Y 2002 Sens.and Act. B 81-2 158

IOP Conf. Series: Materials Science and Engineering **323** (2018) 012004 doi:10.1088/1757-899X/323/1/012004

Production of petroleum bitumen by oxidation of heavy oil residue with sulfur

Ye. Tileuberdi*[1,2], Ye.A. Akkazyn[1,2], Ye.K. Ongarbayev[1,2], Ye.I. Imanbayev[1] and Z.A. Mansurov[1,2]

[1]Institute of Combustion problems, 172, Bogenbai batyr str., Almaty, Kazakhstan
[2]Al-Farabi Kazakh National University, 71, Al-Farabi ave., Almaty, Kazakhstan

*E-mail: Erbol.tileuberdi@kaznu.kz

Abstract. In this paper production of bitumen adding elemental sulfur at oxidation of oil residue are investigated. The objects of research were distilled residue of Karazhanbas crude oil and elemental sulfur. These oil residue characterized by a low output of easy fractions and the high content of tar-asphaltene substances, therefore is the most comprehensible feedstock for producing bitumen. The sulfur is one of the oil product collected in oil extraction regions. Oxidation process of hydrocarbons carried out at temperatures from 180 up to 210 °C without addition of sulfur and with the addition of sulfur (5-10 wt. %) for 4 hours. At 200 °C oxidation of hydrocarbons with 5, 7 and 10 wt.% sulfur within 3-4 h allows receiving paving bitumen on the mark BND 200/300, BND 130/200, BN 90/130 and BN 70/30. Physical and mechanical characteristics of oxidation products with the addition of 5-7 wt. % sulfur corresponds to grade of paving bitumen BND 40/60. At the given temperature oxidized for 2.5-3 h, addition of 10 wt. % sulfur gave the products of oxidation describing on parameters of construction grades of bitumen (BN 90/10).

1. Introduction

Bitumen is a complex combination of organic liquids that is viscous, black and sticky. It is a complex mixture of high boiling point range of compounds and molecules with a relatively low hydrogen-to-carbon ratio. In the United States the terms asphalt and bitumen are synonymous, while in other areas, e.g. in Europe and Asia, both terms have different meanings [1-4].

According to European Standard EN 12597 definition that the bitumen is a virtually in volatile, adhesive and water proofing material derived from crude petroleum, or present in natural asphalt, which is completely or nearly completely soluble in toluene, and very viscous or nearly solid at ambient temperature, whereas asphalt is defined as a mixture of mineral aggregates and bituminous binder. It is not known to present any safety, health or environmental hazard. Meanwhile, the American Society for Testing and Materials (ASTM) defines bitumen as a generic class of amorphous, natural or manufactured, dark colored, cementitious substances composed principally of high molecular mass hydrocarbons, soluble in carbon disulfide. Asphalt also is defined as a cementitious material in which the predominating constituents are bitumens. The terms bituminous and asphaltic then refer to materials that contain or are treated with bitumen or asphalt [2, 5-7].

A plenty of sulfur is taken during development of oil, gas and gas-condensate fields of Kazakhstan. The collected sulfur in oil extraction regions affect to the environment and health of the population.

Solid elemental sulfur is harmless, but there is an allocation in air more than half of hydrogen sulfide dissolved in liquid sulfur during flood of sulfur in sulfuric cards. The sulfuric dust formed at destruction of sulfur, also renders harmful influence and represents the biggest risk. In this connection there was a problem of utilization of sulfur as its open storage renders negative influence on an environment because of interaction with it. Sulfur is bread of chemical industry., a valued commodity in its own right that`s used in some 30000 products. So it`s should be use other economic roads, similar the oil-gas, cool and metals. One of perspective directions of use of elemental sulfur is road construction - extremely material-intensive branch of the construction industry. Here sulfur can be applied as the modifier of paving bitumen to giving of the improved operational characteristics. Also the sulfur estimated cheapness raw material. The USA spends industrial tests of sulfuric knitting sulfeks (60-70 % of sulfur with hydrocarbons and polymers) from 1979. Sulfur with hydrocarbons, can completely replace asphalt in asphalt concrete [8, 9].

Kazakhstan is the ninth largest country in the world, covering an area of 2717300 km^2 and government programs have been agreed for the improvement of infrastructure and road construction across the country [10, 11]. Regarding these reason the bitumen consumption is year by year increasing. The main problem with road building is the poor quality of bitumen used in asphalt-concrete pavements.

In our country the satisfaction of need for a bitumen material occurs due to its import Russia and Iran. Characteristics of imported bitumen completely mismatch climatic conditions of our country, i.e. do not maintain sharp differences of temperatures from -40 up to +40 °C. And also with a view of preservation of quality and technological properties, bitumen is not recommended to be transported on greater distances. In this connection, production of bitumen in various marks on the basis of domestic raw material is an actual problem [12]. In this work is proposed use of sulfur collected in oil extraction regions for production of bitumen.

2. Experimental section

The objects of research were distilled residue of Karazhanbas crude oil and elemental sulfur. Crude oil of Karazhanbas oil field is characterized by a low output of easy fractions and the high content of tar-asphaltene substances, therefore is the most comprehensible feedstock among Kazakhstan crudes for production of bitumen. The sulfur is one of the oil product collected in oil extraction regions.

Oxidation process of hydrocarbons carried out at 3 L of cylindrical reactor made by stainless steel. At process, the heated air blowing given to the raw material from compressor. Oxidation of oil residue carried out at temperatures from 180 up to 210 °C without the addition of sulfur and with the addition of sulfur in quantity from 5 up to 10 wt. %. Through the certain time intervals (2; 2,5; 3; 3,5; 4 h) were selected tests of oxidation product.

Penetration involves the determination of the extent to which a standard needle penetrates a properly prepared sample of bitumen under specified conditions of temperature, load, and time. The unit of penetration is 0.1 mm, which is generally omitted in favor of reporting just the measured number. It was determined by a apparatus Penetrometer PNB-03 in accordance with standard 11501-78. Softening temperature – the temperature at which the bitumen of a relatively solid state in to the liquid state. The softening point was determined by the method of "ring and ball" according to standard 11506-73. The penetration index characterizes the degree of penetration of colloidal bitumen or rejection of his status from a purely viscous. It is determined by the empirical formula.

3. Results and discussion

Modification of oil feedstock by elemental sulfur allows making sulfur-bitumen binders with the improved characteristics under soft conditions of experiment [13]. In this connection, in the given work the opportunity of bitumen production by oxidation of oil residue of Karazhanbas oil field with the addition of elemental sulfur. At first, oxidation process tested at low temperature (at 180 °C) and 5-10 weight percentage of sulfur were added into bitumen. But, results of analyses showed that the almost samples on the physical and mechanical characteristics were poor indicator. At 180 °C necessary values of softening point and penetration under requirements of the standard are reached only at the addition of sulfur in quantity 10 wt. % and time of oxidation 4 h. Therefore, the oxidation temperatures were increased to 200 and 210 °C and results were shown in the below tables.

Table 1. Physical and mechanical characteristics of the oxidation products of Karazhanbas oil residue with the addition of sulfur at 200 °C.

Amount of sulfur added to oil oxidation, mas. %	5			7			10		
Oxidation time, hour	3	3,5	4	3	3,5	4	3	3,5	4
The depth of penetration of the needle into the bitumen at 25 °C, 0.1mm:	295	205	156	206	180	116	27	24	14
Softening point, ° C, not less:	40	42	44	41	45	49	68	78	86
Brittleness temperature, ° C, max:	-	-21	-23	-25	-	-14	-	-	-
Penetration index	2,3	0,9	0,4	0,5	1,5	0,9	1,1	2,3	2,3
Standard accordance	-	BND 200/ 300	BND 130/ 200	BND 200/ 300	-	BN 90/130	-	BN 70/30	-

Table 2. Physical and mechanical characteristics of the oxidation products of oil residue at the temperature 210 °C with sulfur.

Amount of sulfur added to oil oxidation, mas. %	0					10				
Oxidation time, hour	2	2,5	3	3,5	4	2	2,5	3	3,5	4
The depth of penetration of the needle into the bitumen at 25 °C, 0.1 mm:	255	225	155	105	65	10	9	8	3	3
Softening point ring and ball, ° C, not less:	16	23	31	36	41	82	90	98	104	108
Standard accordance			-			-	BN 90/10	BN 90/10	-	
Amount of sulfur added to oil oxidation, mas. %	5					7				
Oxidation time, hour	2	2,5	3	3,5	4	2	2,5	3	3,5	4
The depth of penetration of the needle into the bitumen at 25 °C, 0.1 mm:	276	202	123	86	55	76	56	47	29	24
Softening point ring and ball, ° C, not less:	32	37	40	46	53	44	47	51	57	62
Brittleness temperature, ° C, max:	-	-	-	-	-18	-	-	-12	-18	-10
Penetration index		-1,4	-1,9	-1	-0,2	-1,9	-1,8	1,0	-0,8	0,2
Standard accordance		-			BND 40/60	-		BND 40/60	-	

Apparently from the table 1, the values of softening point and penetration of oxidation products without the addition of sulfur at all investigated temperatures mismatch requirements of the standard. At 200 °C oxidation of hydrocarbons with 5 and 7 wt.% sulfur within 3-4 h allows receiving paving bitumen on the mark BND 200/300, BND 130/200 and BN 90/130. At increase in the content of sulfur up to 10 wt. % physical and mechanical parameters sharply differ and come nearer to norms of construction bitumen (BN70/30).

The technical parameters of oxidation products, which formed at a temperature of 210 °C were significantly changed than the products obtained at a temperature of 200 °C. This occurs due to an increase the oxidation rate when increasing the temperature.

The data on compliance of physical and mechanical characteristics of oxidative products with the standard at the temperature 210 °C are given in table 2. As shown the tabulated dates, the oxidation products which without adding sulfur do not meet the standard on physical and mechanical properties. Physical and mechanical characteristics of oil residue oxidation products with the addition of 5-7 wt. % sulfur corresponds to grade of paving bitumen BND 40/60. At the given temperature oxidized for 2.5-3 h, addition of 10 wt. % sulfur gave the products of oxidation describing on parameters of construction grades of bitumen (BN 90/10).

Physical and mechanical characteristics of oxidized bitumen were compared. The road bitumen's main characteristics are penetration and softening temperatures. They determine the consistency and application areas of bitumen. Dependence or changing of bitumen properties on the oxidation time at the 210 °C are showing in figure 1 and figure 2.

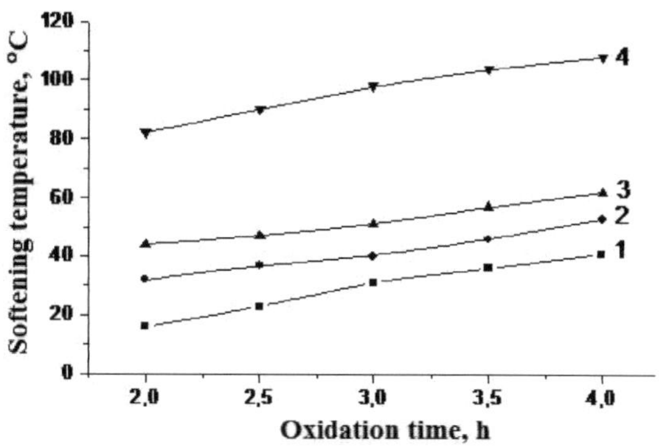

1 – without adding sulfur; with sulfur in the amount: 2 – 5 wt. %, 3 – 7 wt. %, 4 – 10 wt. %

Figure 1. Dependence of penetration on the oxidation time of products at the 210 °C.

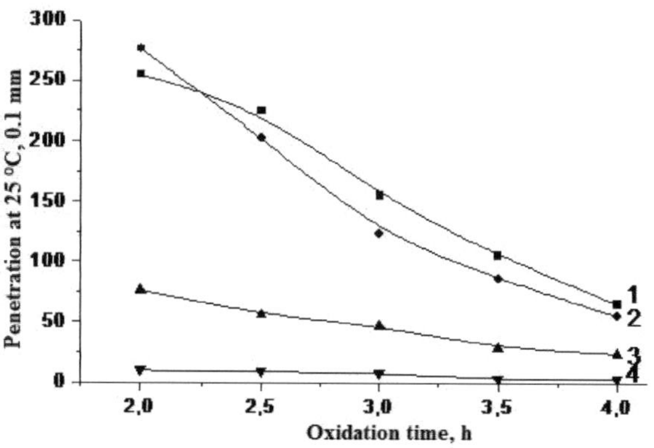

1 – without adding sulfur; with sulfur in the amount: 2 – 5 wt. %, 3 – 7 wt. %, 4 – 10 wt. %

Figure 2. Dependence of softening point on the oxidation time of products at the 210 °C.

When oxidizing oil residue without adding sulfur, within 2 hours the product's softening temperature increased from 16 to 41 ºC, ie the growth is 25 ° C (figure 1). Meanwhile softening temperature sulfur of oxidation products with sulfur had higher. During this time, ST of bitumen adding 5 wt. % of sulfur increased from 32 to 53 ºC and adding 7 wt. % of sulfur increased from 44 to 62 ºC. As seen from figure 2, with increasing content of sulfur from 5 wt. % to 10 wt. % in these processes, the penetration of sulfur bitumen were decreased. Consequently, as the amount of sulfur in the mixture increases, the rate of oxidation of the oil is considerably reduced. The softening temperatures of the 10 % sulfur-added bitumen were increased by 26 °C in 2 hours (from 82 ºC to 108 ºC). In this case, the softening temperature increases rapidly, and the penetration decreases only slightly.

Thus the all results of oxidation process showed that the oxidation products exhibit elasticity, increased softening point, decreased penetration and enhanced strength. It means the sulfur influenced to oxidation of hydrocarbons as oxidizer.

4. Conclusion

Obtained data allow making the conclusion that the addition of elemental sulfur leads to reduction of time of process of oxidation and mitigation of regimes of process. The results of researches have shown, that oxidation of high-viscous oil residue with the addition of elemental sulfur (5-10 wt. %) at low temperatures (200 and 210 °C) for short time (3,5-4 hours) can receive bitumen of satisfactory quality. The resulting of oxidation process exhibit elasticity, increased softening point, decreased brittle point, and enhanced strength. These properties allow sulfur bitumen products to be used both as binders for asphalt concretes and as mastics for pavement repair. Also, it allows partially solving the problem of utilization of sulfur which collected in oil extraction regions.

References

[1] Mansurov Z.A., Ongarbaev E.K., Tuleutaev B.K.: Utilization of Oil Wastes for Production of Road-building materials // Eurasian Chemico-Technological Journal. – 2000. – № 2(2). – P. 161-66.

[2] Songhun Yoon, Sharad Durgashanker Bhatt, Wonkyu Lee, Heung Yeoun Lee, Soon YongJeong, Jin-Ook Baeg, and Chul Wee Le. Separation and characterization of bitumen from Athabasca oil sand // Korean J. Chem. Eng. – 2009. – Vol. **26(1)**. – P. 64-71.

[3] Ye. Tileuberdi, Ye.K. Ongarbaev, Z.A. Mansurov, F. Behrendt. Nanostructure of bitumen produced from heavy oil. – Almaty: Qazaq University, 2017. –P. 164.

[4] Asphalt and bitumen. Encyclopedia of chemistry (Ullmann's). –2012. –Vol. **4**. – P. 273-94. DOI: 10.1002/14356007.a03_169.pub2

[5] Tileuberdi E., Ongarbayev Y.K., Imanbayev Y.I., Tulepov M.I., Mansurov Z.A., Tuleutaev B.K. Production of Rubberized Bitumen by oxidation of black oil // Material Science and Environmental Engineering. – CRC Press, 2016. – P. 183-86.

[6] Recommendations for the Safe Handling and Storage of Bitumen Emulsions. http://www.rea.org.uk. 16.11.2014.

[7] Imanbayev Y., Ongarbayev Y., Tileuberdi Y., Krivtsov E., Golovko A., Mansurov Z. High temperature transformation of tar-asphaltene components of oil sand bitumen // Journal of the Serbian Chemical Society. – 2017. – Vol. **82**, – No. 9. – P. 1063-73.

[8] Ludwig Allen, C. Ind. and Eng. Chem. Prod. Res. and Development. 21(1):76 (1982).

[9] Gureev, A.A., Larina, N.M., Abi-Fadel, Yu., and Fedorov, A.A. Modification of the properties of paving asphalts by treatment of vacuum resid with sulfur. Chemistry and Technology of Fuels and Oils. 38 (5):325 (2002).

[10] Akkazin E.A. Production of bitumen by oxidation of Karazhanbas crude oil and oil residue: Authorref. ... cand. chem. sci.: 02.00.13. – Almaty: Areket-Print, 2009. – 22 p.

[11] Oil Sands in Kazakhstan. http://www.inogenet.com. 20.07.2012.

[12] Tileuberdi E., Ongarbayev Ye.K., Behrendt F., Schneider I., Imanbayev Ye.I., Tuleutayev B.K., Doszhanov Ye.O., Mansurov Z.A. Possibilities of preparation asphalt concrete by oil sands of Kazakhstan // KazNU Bulletin. Chemical series. – 2012. – № 4 (68). – P. 9-13.

[13] Mansurov, Z.A., Ongarbaev, E.K. and Tuleutaev, B.K. Contamination of soil by crude oil and drilling muds. Use of wastes by production of road construction materials. Chemistry and technology of fuels and oils. 37 (6):441 (2001).

IOP Conf. Series: Materials Science and Engineering **323** (2018) 012005 doi:10.1088/1757-899X/323/1/012005

NiCo as catalyst for magnetically induced dry reforming of methane

F. Varsano[1]*, M. Bellusci[1], A. Provini[2] and M. Petrecca[3]
[1]ENEA C.R. Casaccia, Via Anguillarese 301, 00123 Roma, Italy
[2]Università degli Studi di Genova,Via Dodecaneso 31, 16146 Genova, Italy
[3]Dip. Chimica-Università di Firenze,Via della Lastruccia3, 50019 Sesto Fiorentino (FI), Italy

*Email: francesca.varsano@enea.it

Abstract. In this paper we report the activation of the dry reforming reaction by induction heating of a NiCo alloy. The catalyst plays a double role, serving both as a promoter for the reforming reaction and producing the heat induced by dissipation of the electromagnetic energy. The elevated temperatures imposed by the reforming reaction suggest the choice of an alloy with a Curie temperature >800°C. In this respect Ni:Co ratio 60:40 was chosen. Alloy active sites for CH_4 and CO_2 activation are created by a mechanochemical treatment of the alloy that increases solid-state defects.
The catalyst has been successfully tested in a continuous-flow reactor working under atmospheric pressure. Methane conversion and hydrogen production yields have been measured as a function of the applied magnetic field, reactant flow rate and time on stream.

1. Introduction

The possibility of utilizing magnetic materials in catalytic processes has been recently accomplished in the field of magnetically assisted fluidized bed reactors [1] and magnetic separations [2]. Recently, the opportunity of converting electromagnetic energy into heat by means of magnetic materials has been proposed in the field of catalysis [3]. In this respect, energy is supplied by induction under radiofrequency field in a fast and efficient way to a properly responding catalytic system generally containing both a magnetic and catalytic component. In such a way, heat is provided directly by the catalytic bed itself that selectively absorbs electromagnetic energy. Heat is promptly generated on the chemically active site and in principle, by a precise design of the magnetic material, it should be possible to supply exclusively the energy required to drive the chemical process, avoiding unnecessary heating (and dissipation) of the whole reactor and increasing the overall efficiency of the process.

The catalytic activity is usually performed by a different non magnetic component that it is either supported on the magnetic material or in close proximity.

So far, such approach has been successfully applied to "low temperature" chemical processes such as the direct synthesis of amide from amine and carboxylic acid [4] at 170°C, the Claisen rearrangement (200°C) and many areas of organic synthesis (90-140°C) [5, 6]. Specific catalysts with optimized magnetic properties have been synthesized in order to adjust the reachable temperature and heating efficiency. Meffre et al. [7] reported the synthesis of complex nano-materials (Fe@FeCo, Fe@Ru) displaying high specific absorption rate (SAR) under external magnetic field. The nano-particles heated by induction were able to catalyze CO hydrogenation leading to Fischer-Tropsch syntheses (T>200°C). Magnetic zeolites (TiO_2-coated nickel ferrite particles coated with a layer of ZSM-5 zeolite crystals) were successfully tested for the isomerization of citronellal at 350°C [8]. Bordet at al. designed iron carbide nanoparticle with remarkable heating properties that when associated with catalytic metals (Ni, Ru) were able to efficiently catalyze CO_2 hydrogenation in a continuous-flow reactor (~350°C) [9].

This field of investigation is new and large improvement is expected in the design of specific catalysts with enhanced magnetic properties. Presently the achievement of high temperatures by means of magnetic nanoparticles seems challenging. To overcome this problem NiCo alloy has been synthesized by arc melting and successively grinded to the micrometer range by a mechanochemical process. Here we report on the catalytic properties of the material heated by induction under r.f. field toward a high temperature endothermic reaction, the dry methane reforming. The reaction is highly endothermic (H_{298K}=247kJmol^{-1}) and requires operating temperature in the range 800-1000°C to accomplish high conversion of CH_4 and CO_2 to H_2 and CO. The most commonly used catalysts, apart from noble metal, are nickel based.

2. Material properties and applied methodologies

A NiCo alloy with atomic ratio (60:40) has been synthesized by arc melting (Edmund Buler GmbH). The alloy has been fragmented into powder by a mechanochemical treatment (20min) using a vibrating cup mill inside a tempered steel grinding set. The material has been characterized by X Ray Diffraction (XRD) in the angular range 4° < 2θ< 40°, using a Seifert Pad VI apparatus equipped with Mo Kα radiation and a LiF monochromator on the diffracted beam. The resulting spectrum is reported in figure 1. It shows a single phase of NiCo alloy with an average crystallite dimension of ~15nm (Debye-Sherrer). Scanning electron microscopy (SEM) images were acquired by using a high-resolution microscope (HRSEM LEO 130). In figure 2 a SEM picture of NiCo powder is reported showing particles with dimensions in the micrometer range.

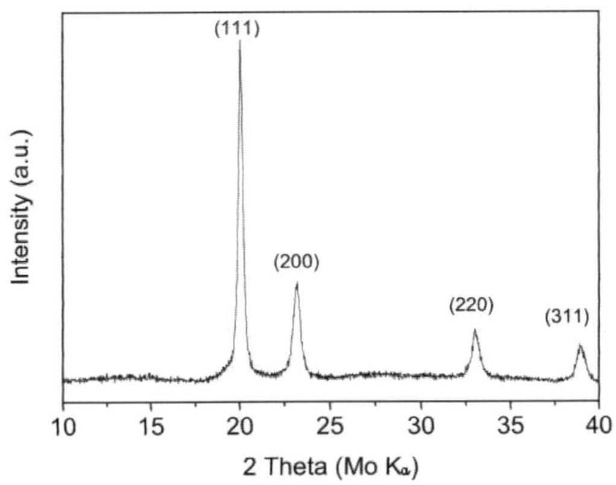

Figure 1. XRD spectrum of NiCo alloy.

Figure 2. SEM image of NiCo powder as prepared.

Magnetic measurements have been carried out on tightly packed, randomly oriented powder using a SQUID magnetometer (Quantum Design MPMS). Static magnetic properties of the as-prepared powder have been investigated by recording the hysteresis loop at room and low temperature (5 K), shown in figure 3. As expected, the sample displays very low coercive field (H_C), 70 Oe and 115 Oe for 300 K and 5 K, respectively. Saturation of the magnetization is already obtained at 10 kOe, and the saturation values slightly decrease on increasing temperature from 52 emu/g at 5 K to 50 emu/g at room temperature. Both the low coercivity and easiness of saturation agree with the soft magnetic properties expected for a NiCo alloy.

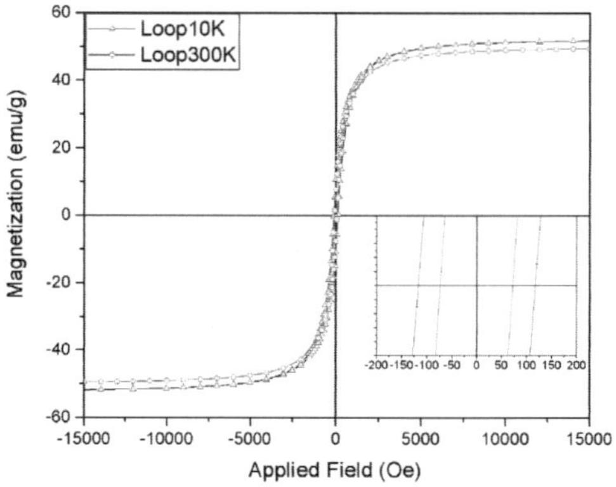

Figure 3. Hysteresis loops of NiCo alloy powder at room temperature (red curve) and 5K (blue curve). In the inset a magnification of the low field region is shown.

The capability of the NiCo alloy powder to act as a heat mediator has been evaluated by recording the temperature kinetics upon application of an alternating magnetic field with frequency f = 183 kHz and variable amplitude, H_0 from 3.0 to 17.0 kA/m. The sample was prepared pressing the powder into a pellet (8 mm diameter, 1mm thick) and the temperature was monitored using and optical fibre thermometer in contact with the surface of the pellet. The resulting temperature curve as a function of time (s) is showed in figure 4. Extrapolating the initial slope of this curve from a linear fit we estimated a temperature increment per second normalized to the sample mass of 66 °C/s•g, which, considering the heat capacity of the bulk alloy (0.468 J/g°C^{-1} for Ni0.6Co0.4) corresponds to a dissipated power of 9.3 W/g. It is worth to note that the dissipated power scales linearly with the applied field.

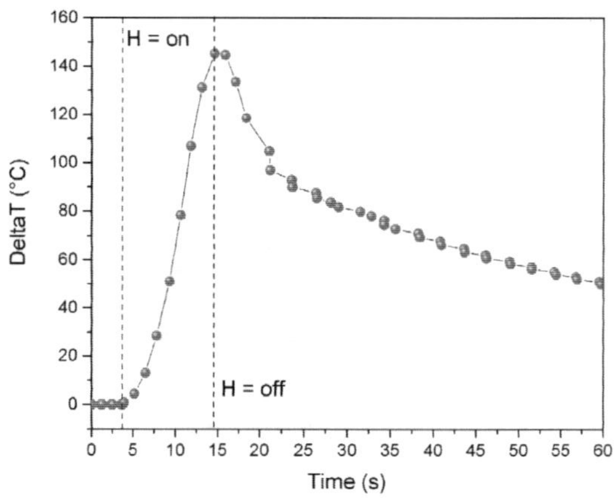

Figure 4. Temperature kinetics of the powder under the application of an alternating field of H_0 = 17 kA/m and f = 183 kHz. The starting temperature was 22°C.

Functional tests have been performed utilizing the quartz reactor and gas distribution lines of a Temperature Programmed Desorption Analyzer-TPD (Micromeritics Autochem 2920). Catalyst powders were compacted (5ton/cm^2) into a disc (8mm diameter, 1mm thick) and placed inside a U quartz tube (figure 5). Before its utilization in the process conditions, the sample was reduced in He-H_2 (3%vol), 50ml/min flow, to clean the oxidized surface through a thermal treatment in TPD at

350°C (10 minutes). An Ambrell EasyHeat 2.4 apparatus has been utilized to heat the catalyst. The inductor is placed around the quartz tube as shown in figure 6. In figure 7 the power supplied to the coil to generate the magnetic field is displayed.

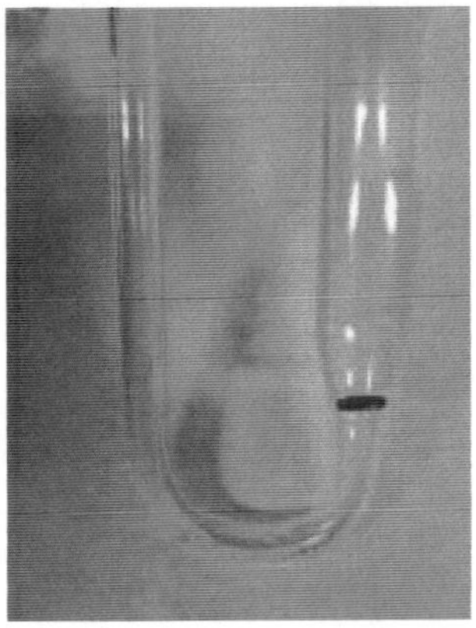

Figure 5. NiCo powders are compacted into a pellet (d=8mm, h=1mm) and placed in the quartz reactor. Reactive gases flow through the disc.

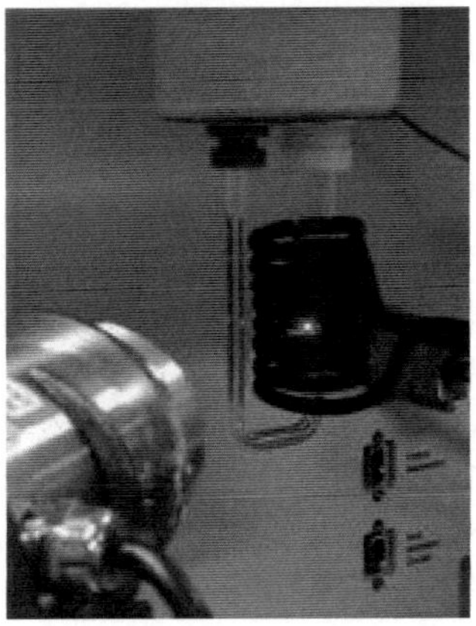

Figure 6. Inductor coil (d= 2.5cm, h=4.5cm) surrounding the reactor tube. A pyrometer provides an estimation of the catalyst surface temperature.

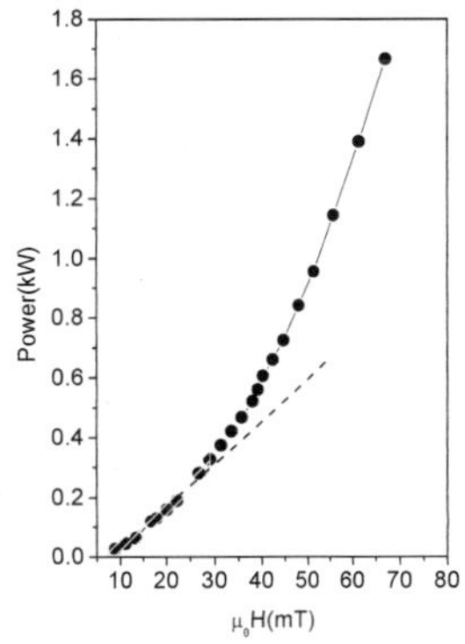

Figure 7. Power consumption vs. generated magnetic field for the coil utilized in this work. (f=190KHz). The linear response of the coil is highlighted.

The experimental set-up includes an optic pyrometer (IRTech Radiamatic IR20CF2150) to estimate the catalyst surface temperature. Such value is only an approximation of the real temperature since the emissivity of the catalyst is not known. GC analysis (Agilent GC490 equipped with two independent columns MS5A and PoraplotU) is utilized to determine the composition of the gas leaving the reactor.

Samples are heated by induction at 190kHz under argon flow until the temperature reaches a stable temperature (few minutes), then the gas is switched to a reactive mixture composed of 50ml/min Ar-CH_4 (10%) and 5ml/min CO_2.

3. Catalytic activity

The NiCo catalyst activity toward the dry reforming reaction was firstly investigated in the TPD reactor by conventional heating and results are reported in figure 8. Methane conversion and hydrogen production yield exceeding 70% were observed at 850°C. The measured conversion values are considered a reference to which results obtained by induction heating are compared.

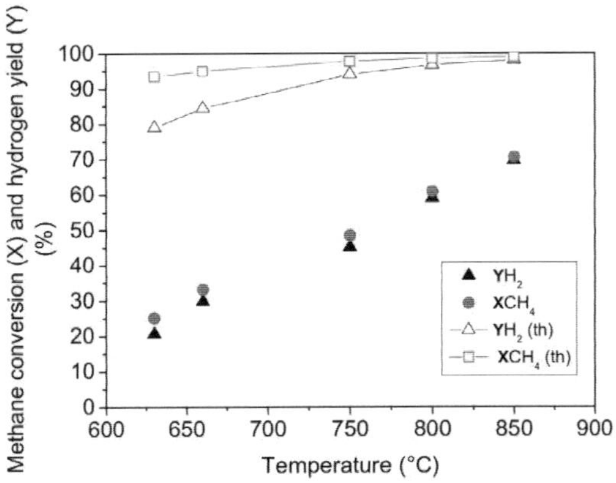

Figure 8. Methane conversion $X_{CH_4} = \frac{[CH_4]in-[CH_4]out}{[CH_4]in}.100$ and hydrogen yield $Y_{H_2} = \frac{[H_2]out}{2[CH_4]in}.100$ measured when NiCo catalyst is conventionally heated at different temperatures. Theoretical expected values, calculated on the basis of thermodynamic equilibrium data (HSC Chemistry 6.1 software) taking into consideration multiple equilibria, are reported.

The NiCo activity was measured heating the catalyst by induction for magnetic field amplitudes ranging from 26 to 32mT (limited to the coil linear response range in figure 7). In figure 9 the methane conversion (X_{CH4}) and hydrogen yield (Y_{H2}) are reported as a function of the applied magnetic field. When compared to measurements performed by conventional heating, the catalyst shows a better performance suggesting that NiCo surface is higher than 850°C. In figure 10 the effect of the gas feed on the catalyst activity is reported, keeping CH_4:CO_2 ratio 1:1. Conversion data versus time on stream are reported in figure 11. These data, a preliminary investigation on catalyst stability, are very promising considering that the metallic catalyst is not dispersed or supported on an inert substrate and that annealing phenomena could easily deactivate it.

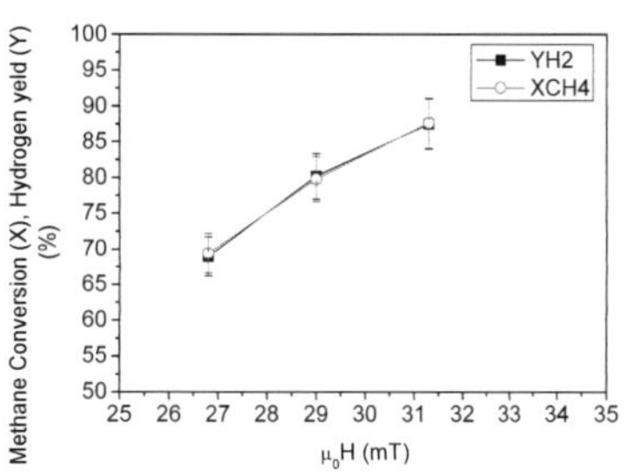

Figure 9. Methane conversion (X_{CH4}) and hydrogen yield (Y_{H2}) as a function of magnetic field amplitude. GHSV=9330h^{-1}.

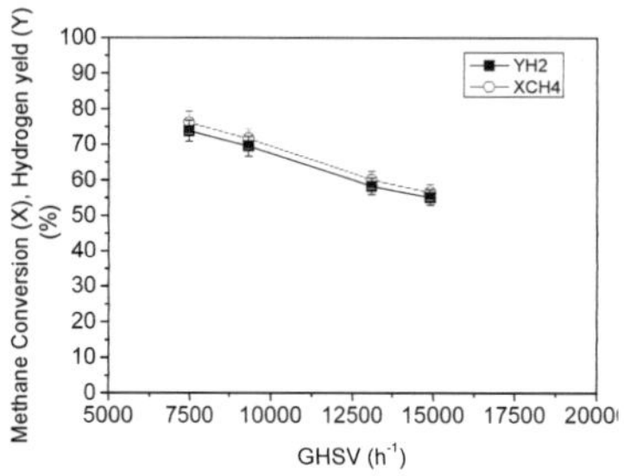

Figure 10. Effect of the gas feed on methane conversion (X_{CH4}) and hydrogen yield (Y_{H2}). Magnetic field amplitude 27mT, 0.32kW.

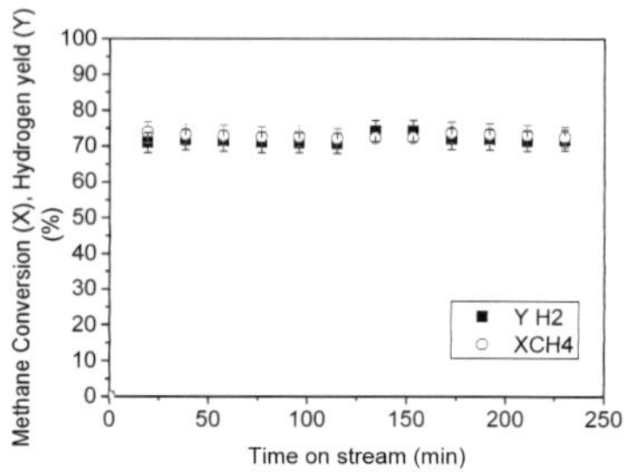

Figure 11. Catalytic results as a function of time on stream. Preliminary test of the stability of the catalyst. Magnetic field amplitude 27mT, 0.32kW, GHSV=9330h^{-1}.

The mass balance between gases feed and outlet shows a carbon deficit (0.5-3%). Preliminary SEM observation does not evidence carbon deposits on the powder and a longer time on stream operation is needed to clarify this indication.

4. Conclusion

A NiCo alloy catalyst is proposed to carry out the dry reforming of methane supplying the required energy by a radiofrequency alternating electromagnetic field. Catalyst powders are heated by induction to very high temperatures (>850°C). The developed material has been utilized at the same time as a catalyst and heating agent to perform dry-methane reforming with conversion values up to 90%. Preliminary tests have been performed to evaluate the effect of gas feed and the stability of the developed catalyst. Reported evidences open new opportunities in the research of magnetic materials for industrial chemical processes.

References

[1] Baoning Z, Xiangkun , Xuhong M, Ziaoxin Z, 2013 Magnetically stabilized bed reactors *Chinese J. Catalysis* **34** 61-68

[2] Polshettiwar V, Luque R, Fihri A, Zhu H, Bouhrara M, Basset JM 2011 Magnetically recoverable nanocatalysts *Chem. Rev.* **111** 3036-75

[3] Houlding TK, Rebrov EV, 2012 Application of alternative energy forms in catalytic reactor engineering, *Green Process Synth* **1** 19-31

[4] Houlding TK, Gao P, Degirmenci V, Tchabanencko K, Rebrov EV, 2015 Mechanochemical synthesis of TiO_2/$NiFe_2O_4$ magnetic catalysts for operation under RF field *Mater. Sci. Engineer. B* **19**, 175-80

[5] Ceylan S, Friese C, Lammel C, Mazac K, Kirschning A 2008 Inductive Heating for Organic Synthesis by Using Functionalized Magnetic Nanoparticles Inside Microreactors, *Angew.*

IOP Conf. Series: Materials Science and Engineering **323** (2018) 012005 doi:10.1088/1757-899X/323/1/012005

Chem. Int. Ed. **47** 8950-53

[6] Kirschning A, Kupracz L, Hartwig J, 2012 New synthetic opportunities in miniaturized flow reactors with inductive heating, *Chem. Lett.* **41** 562-70

[7] Meffre A, Mehdaoui B, Connord V, Carrey J, Fazzini PF, Lachaize S, Respaud M, Chaudret B, 2105 Complex nano-object displaying both magnetic and catalystic properties : a proof of concept for magnetically induced heterogeneous catalysis, *Nano Lett.* **15** 3241-48

[8] Garcıa-Aguilar J, Fernandez-Garcıa J, Rebrov EV, Lees MR, Gao P, Cazorla-Amoros, A. Berenguer-Murcia, Magnetic zeolites: novel nanoreactors through radiofrequency heating, *Chem. Commun.*, 2017, 53, 4262-65

[9] Bordet A, Lacroix LM, Fazzini PF, Carrey J, Soulantica K, Chaudret B, 2016 Magnetically induced continuous CO_2 hydrogenation using composite iron carbide nanoparticles of exceptionally high heating power, *Angew. Chem. Int. Ed.* **55** 15894-98

IOP Conf. Series: Materials Science and Engineering **323** (2018) 012006 doi:10.1088/1757-899X/323/1/012006

Potential adsorption of methylene blue from aqueous solution using green macroalgae *Posidonia oceanica.*

F-N Allouche and N Yassaa

Centre de Développement des Energies Renouvelables, B.P. 62, Route de l'Observatoire, Bouzaréah 16340-Algiers, Algeria.

*E-mail: n.allouche@cder.dz

Abstract. The use of inexpensive biological materials, such as marine algae for removing dyes from contaminated industrial effluents appears as a potential alternative method. The aim of this study is to investigate the aptitude of marine macroalgae Posidonia Oceanica local biomass abundant on the coasts of Algeria for selective sorption of methylene blue (MB) from an aqueous solution in batch experiments at 20 °C. A maximum percentage removal of Posidonia oceanica occurs at pH 5. Equilibrium isotherm data were analyzed using the Langmuir and the Freundlich isotherms. The adsorption equilibrium of methylene blue was best describe by Langmuir model than the Freundlich model. The maximum sorption capacity was 357 mgg^{-1} at pH 5. The sorption data were very well described by the pseudo-second-order model.
Keywords: Posidonia oceanica, Methylene blue (MB), Biosorption, Isotherm Equilibrium, Kinetics; Modelling.

1. Introduction

In recent years, much attention has focused on various natural solids material, which are able to remove pollution from contaminated wastewater at low cost. *Posidinia oceanica* was an important part of the marine ecosystem and can be developed for number of application. *Posidonia oceanica* are among the most macroalgae in studies of dyes pollution due to their ubiquitous environment and readily available in Mediterranean coast. The marine green macroalgae *Posidonia oceanica* are a renewable natural biomass and are very abundant in the Algerian littoral. Considerable research has been carried out by numbers of studies in Posidonia oceanica biosorbent, for their ability to sequester dye from aqueous solutions [1-5]. This study is intended to highlight on the possibility to use *Posidonia oceanica* as new material to remove and adsorb hazardous pollutants from industrial effluents. Nowadays, it's well known that the pollution due to synthetic dyes is considered as a serious public concern. An untreated dyes cause chemical and biological changes and lead a number of environmental and health hazards. Nearly, 10.000 dyes available commercially and different group of dyes can be discerned [6]. Consequently, approximately 1000000 kg/year of dyes are discharged into waste streams by the textile industry [7]. Most of the industrial activities use synthetic dyestuffs in a great number of operations, they consumes large quantities of water and they generate a considerable amount of colored wastewater. They can be highly toxic and carcinogenic to human health. Methylene bleu (MB) 3,7-bis(dimethylamino)- phenazathionium chloride (a thiazine cationic dye) (figure 1) is commonly used for coloring paper, temporary hair colorant, dyeing cotton, wools and so on [8]. Several physicochemical wastewater treatment techniques were developed for dyes recovery from industrial effluents including physical and chemical processes such as precipitation, sedimentation, ultrafiltration, flotation, color irradiation, ozonation and coagulation. For the above-mentioned technologies are sometimes inappropriate especially for low-concentration solutions due to economic, technical or environmental reasons. Adsorption is an effective alternative process for treatment of contaminated wastewater in terms of simplicity of design and operation, and insensitivity of toxic substances [9]. Considerable amount of researches have been done on the development of effective, low cost and easily available alternative biosorbents [10-16]. The main objective of this research is to study the ability of Posidonia for removing methylene blue (MB) from aqueous solutions. A laboratory study was conducted to investigate the feasibility of the use of *Posidonia oceanica* as biosorbent on batch experiments. The effect of pH value and various concentrations on biosorption capacity of *Posidonia oceanica* were evaluate. Biosorption kinetics and isotherms equilibrium were examined.

2. Materials and methods

2.1. Preparation of the biomass

Fresh samples of the local *P.oceanica* biomass were harvested from the west coast of Algiers. Sample preparation consist on preliminary cleaning, the raw material was rinsed with tap water to remove impurities (sand and salt of the surface), and then it was rinsed with distilled water and dried on sun light. Before use, the biomass was cuts into segments of 2-4 cm. A pre-treatment of biomass was carried with 0.1 mol HCL for 1 h under stirring, and then washed thoroughly with distilled water at several time to remove excess of salts from the biomass and finally dried in oven at 60°C for 24 h. Then, they were ground in laboratory blender and stored at ambient temperature.

2.2. Chemicals

All chemicals used in this study were of analytical grade and solutions were prepared using deionized and distilled water. Stock solution of methylene blue was prepared by dissolving $C_{16}H_{18}ClN_3S$ x H_2OX in water figure 1. The concentration of unadsorbed methylene blue dyes in the aqueous solution were measured colorimetrically using a spectrophotometer. The absorbance of the color of methylene blue was read at 665 nm.

2.3. Batch biosorption experiments

2.3.1. Equilibrium studies

The performance of *P.oceanica* is described through the batch experiment. Firstly, adsorption isotherms, 100 mg of biosorbent were mixed with 100 mL of methylene blue solution during 24 h, at 20 °C and 150 rpm stirring rate, at the optimum pH 5. The pH was controlled using molar solutions HCl and NaOH. After this, aliquots were removed from the supernatant and filtered through membranes using 1-2 μm pore size filtration membrane. Filtrates were analyzed by spectrometry (S), using a UV-vis spectrometer (Biomate spectrometer-Thermo electron corporation, Model TM3). The amount of adsorbed by biosorbent at equilibrium was calculated from the mass balance of the equation as given follows:

$$q = \frac{V(C_0 - C_e)}{m} \tag{1}$$

where q is the adsorption amount at equilibrium (mg/g), C_o and C_e are metal concentrations in the initial solution and after adsorption (mg L−1), respectively; V is the volume of solution (L) and W is the mass of biosorbent used (g). For all experiments, the MB concentration was determined by an UV-visible spectrophotometer at wavelength corresponding to the maximum absorbance of the dye solution (λmax=665nm).

2.3.2. Kinetic experiments

Kinetics experiments were carried out to determine equilibrium time for biosorbent, by adding a known amount of *P.oceanica* and methylene blue solution. Samples were taken at fixed time intervals.

IOP Conf. Series: Materials Science and Engineering **323** (2018) 012006 doi:10.1088/1757-899X/323/1/012006

The pH and temperature of the solution were maintained at pH 5 and 20 °C respectively. The flasks containing 500 ml of methylene blue solution of known initial concentration with 1 gL^{-1} of algal biomass and stirrer at constant agitation speed 200 rpm to ensure equilibrium was reached.

3. Results and discussion

3.1 Influence of pH

The pH of aqueous solution plays an important role in the adsorption of dyes onto biomass [17]. The pH affects significantly the adsorption capacities of the Methylene Blue onto the biomass. Figure 2. shows the influence of pH on the sorption of methylene blue using P.oceanica for pH ranging between 2 and 7. The adsorption capacity of methylene blue increases sharply when pH is below 3.0, and then the adsorption becomes to reach the equilibrium state at pH 5. The pretreatment of the biomass with HCl allows reducing the variation of pH during dyes sorption taking into account the characteristics of the biosorbent. This is an important parameter for stabilizing the material, and reducing the possible occurrence of local phenomena of dyes micro-precipitation and for comparison of sorption performance.

Hamdaoui [10] showed that adsorption of Methylene Blue on sawdust and crushed brick increased by increasing pH (until a value of 9). For pH lower than 5 both adsorbents were positively charged: in this case, the adsorption decreased because methylene blue is a cationic dye. At pH 2 and 5, a competition occurs between protonated MB and also there exists an electrostatic repulsion between the positively charged surface of biomass and BM [18].

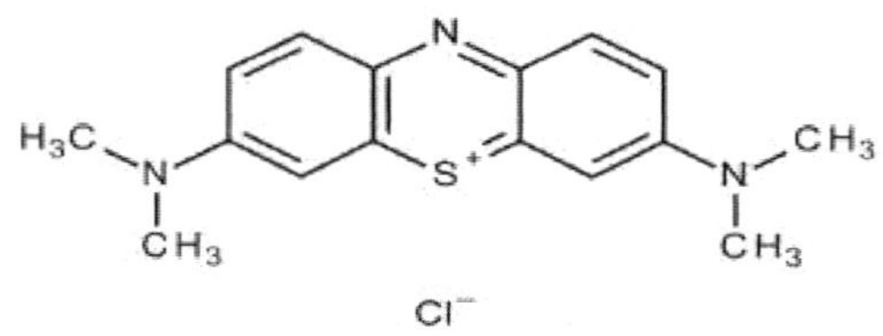

Figure 1. Chemical Structure of Methylene bleu (MB).

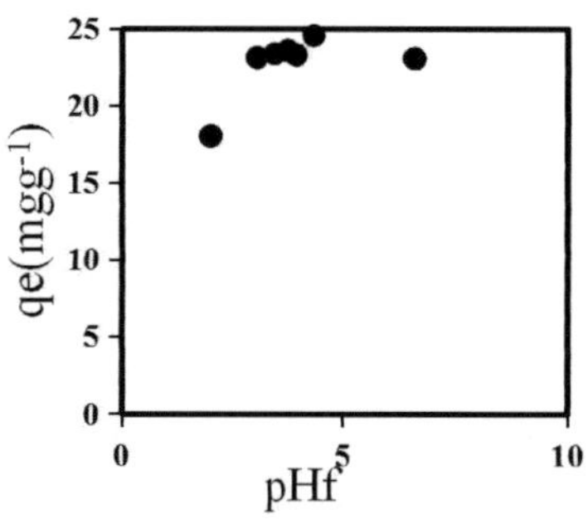

Figure 2. Influence of pH of BM sorption using Posidonia oceanica.

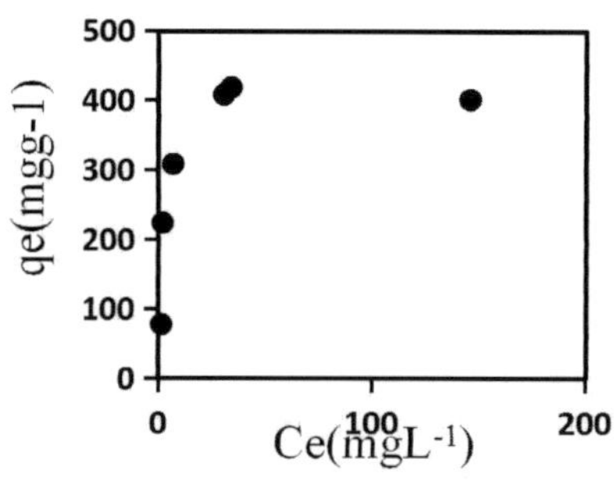

Figure 3. Experimental equilibrium data for modeling with Langmuir isotherm.

3.2. Sorption isotherm

Figure 3 shows the sorption isotherms of the methylene blue biosorption using dried P.oceanica. The isotherms were drawn based on equilibrium concentration. All experiments were carried out initial methylene blue concentration in the range 5-100 ppm and the equilibrium pH was 5. There are numerous reports on dyes biosorption with *P.oceanica* [2-4, 13]. In this study, two models Langmuir (1916) [19] and Freundlich (1906) [20] were commonly used and the appropriate model was identified to better understand the dye–podidonia interaction. From the best correlation obtained by using Langmuir model [19], it can be concluded that the monolayer adsorption is more suitable to explain the adsorption of (MB) and indicates at the same time that chemisorption might also be involved in the sorption process. The maximum sorption capacity reached 357 mgg^{-1} at pH 5. The affinity coefficient (b coefficient) was close to 1.07 Lmg^{-1} at pH 5(table1).

Table 1. Maximum adsorption capacity (q max) of methylene blue by *P.oceanica*.

Dye	Langmuir constant		
Methylene blue	q(mgg^{-1})	b	R2
	357	1.07	0.995

The data demonstrated that the biosorbent in this work had relatively high capacities of the methylene blue dyes. Comparative and competitive biosorbents for methylene blue sorption are reported in table 2. *Posidonia oceanica* can be compete with others adsorbents to remove methylene blue such as zeolite [21], rice husk [22], *Caulerpa cemosa* [23], clay [24], and commercial activated carbon [25, 26]. It can be seen that the qm value varies significantly for different biosorbents and that by comparison, the posidonia indicates a good capacity to adsorb (MB) ions from aqueous solutions.

Table 2. Comparison of methylene blue sorption for different sorbents.

Biosorbent	qm or qmax (mgg^{-1})	Reference
Activated charcoal	25.25	[12]
Posidonia oceanica	5.56	[14]
Crosslinked chitosan /bentonite composite	142.86	[15]
Zeolite	10.82	[21]
Rice husk	40.58	[22]
Caulerpa cemosa Var.cylindracea	5.23	[23]
Clay	6.3	[24]
Commercial activated carbon	930.3	[25]
Commercial activated carbon	200	[26]
Posidonia oceanica	357	In this study

3.3. Adsorption Kinetics

3.3.1. Effect of initial dye concentration

The influence of concentration was studied at room temperature by fixing the solution pH and varying the initial concentration in the range of 25-50-100 mg/L for MB dye. Figure .4. shows the effect of contact time of methylene blue adsorption. It can be noticed that the sorption processes at different concentrations are fast in the time within 30 min. In order to describe adsorption kinetics, the two most commonly used kinetic models of pseudo first-order [27], pseudo second-order kinetic [28] were used to test the experiment data, respectively.

The first-order rate model may be represented in Eq. (2):

$$\mathrm{Ln}\,(qe - qt) = \ln qe - k_1 t \quad (2)$$

where qt and qe are the amount of methylene blue adsorbed (mgg^{-1}) at time t (min) and at equilibrium, respectively, and k_1 is the rate constant of the pseudo-first-order adsorption process (min^{-1}).The pseudo-second-order model is expressed in Eq. (3):

$$\frac{1}{qt} = \frac{1}{k_2 q_e^2} + \frac{1}{q_e} t \quad (3)$$

where k_2 is the constant of pseudo-second-order rate ($g\ mg^{-1}\ min^{-1}$). On the basis of the experimental data of q_t, the equilibrium adsorption capacity qe and k_2 can be determined from the slope and intercept of the plots of t/q_t against t.

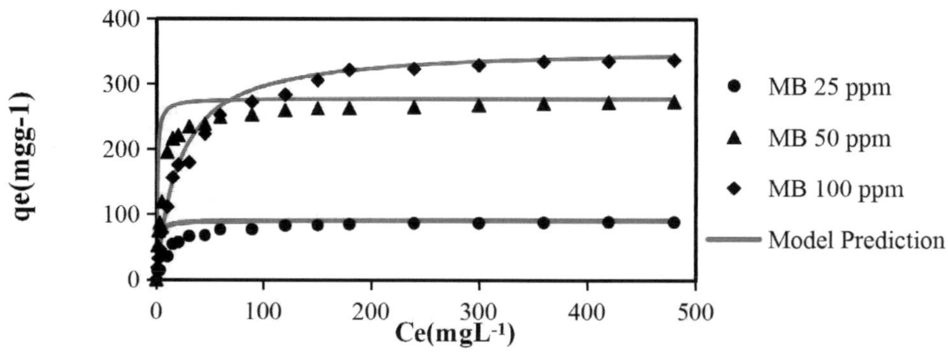

Figure 4. Influence of initial dye concentration (C0, mgMBL^{-1}) on MB uptake kinetics (pH 5; SD: 0.1 gL^{-1}; C0: 50 mg MBL^{-1}; PS: G3; curves represent the modeling of uptake kinetics with the PSORE).

Figure 4 presented the effect of concentration on the dye adsorption capacities of MB dyes on the P.oceanica. An increase in the concentration led to an increase in initial adsorption rate. Based on the obtained R2, of the linear forms of different models, it is observed that pseudo-second-order rate model describes the mechanism of sorption of MB on biomass better than pseudo first-order model.

4. Conclusion

Methylene blue (MB) was selected as a model dye in order to evaluate the capacity of the P.oceanica to remove dyes from aqueous solutions. The pH optimum is found close to pH 5. The biomass P.oceanica can be used as good as new biosorbent for the biosorption of MB. Maximum sorption capacities 357 mgg^{-1} at pH 5 still comparable to the level reached by some biosorbent in the literature. Kinetic experiment can be well represented by the pseudo-second-order kinetic model. The adsorption isotherms could be well fitted by the Langmuir equation confirming that adsorption of (MB) onto P.oceanica ocured by chemisorption. P.oceanica can be developed as cost effective adsorbent from textile and other wastewater.

References

[1] I. Guezguez, S. Dridi-Dhaouadi, F. Mhenni., (2009). Sorption of Yellow 59 on Posidonia oceanica, a non-conventional biosorbent: Comparison with activated carbons, industrial crops and products, 29 197–204.

[2] M. Ulas Dural, Levent Cavas, Sergios K. Papageorgiou, Fotis K. Katsaros., (2011). Methylene blue adsorption on activated carbon prepared from Posidonia oceanica (L.) dead leaves: Kinetics and equilibrium studies. Chemical Engineering Journal, 168 77–85.

[3] S. Cengiz, Fatmagul Tanrikulu, Sevil Aksu., (2012). An Alternative source of adsorbent for the removal of dyes from textile waters: Posidonia oceanica (L.). Chemical Engineering Journal, 189–190 32–40.

[4] N. Ben Douissa, Latifa Bergaoui, Samar Mansouri, Ramzi Khiari, Mohamed Farouk Mhenni., (2013). Macroscopic and microscopic studies of methylene blue sorption onto extracted celluloses from Posidonia oceanica. Industrial Crops and Products, 45 106–13.

[5] N. Ben Douissa, Sonia Dridi-Dhaouadi, Mohamed Farouk Mhenni., (2014). Study of antagonistic effect in the simultaneous removal of two textile dyes onto cellulose extracted from Posidonia oceanica using derivative spectrophotometric method. Journal of Water Process Engineering, 2 1–9.

[6] H. Zollinger., (1987). Colour Chemistry—Synthesis Properties and Applications of Organic Dyes and Pigments, WILEY-VCH, New York, pp, 92-100.

[7] Wong, Y.C., Szeto, Y.S., Cheung, W.H., McKay, G., (2004). Adsorption of acid dyes on chitosan – equilibrium isotherm analyses. Process Biochemistry, 39 695–704.

[8] Haice Karaer, Ismet Kay., (2016). Synthesis, characterization of magnetic chitosan/active

IOP Conf. Series: Materials Science and Engineering **323** (2018) 012006 doi:10.1088/1757-899X/323/1/012006

charcoal composite and using at the adsorption of methylene blue and reactive blue 4. Micropoous and Mesoporous Materaials, 232 26-38.
[9] Tong DS, Zhou CH, Lu Y, Yu H, Zhang GF, Yu WH., (2010). Adsorption of acid red G dye on octadecyl trimethylammonium montmorillonite. Appl Clay Sci, 50 427–31.
[10] Hamdaoui O., (2006). Batch study of liquid-phase adsorption of methylene blue using cedar sawdust and crushed brick. Journal of hazardous Material B, 135 264-73.
[11] Hamdaoui O.M. Chiha., (2007). Removal of methylene blue from aqueous solutions by wheat bran, ActaChim. Slov.54 407-418.
[12] M.J. Iqbal, M.N. Ashiq., (2007). Adsorption of dyes from aqueous solutions on activated charcoal, Journal of hazardous Material B. 139 57-66.
[13] M C., Ncibi, Borhane Mahjoub, Mongi Seffen., (2008).Investigation of the sorption mechanisms of metal-complexed dye onto Posidonia oceanica (L.) fibres through kinetic modelling analysis, Bioresource Technology. 99 5582–89.
[14] M.C. Ncibi, B. Mahjoub, M. Seffen, F. Brouers, S. Gaspard., (2009). Sorption dynamic investigation of chromium (VI) onto Posidonia oceanica fibres: Kinetic modelling using new generalized fractal equation. Biochemical Engineering Journal, 46 141–46.
[15] Buhut H., karear., (2015). Adsorption of methylene blue from aqueous solution by crosslinked chitosan /bentonite composite, J. Dispersion Sci.Technol, 36 61–67.
[16] Z. Aksu., S. Eyda., F.Gonen., (2007). Continuous fixed bed biosorption of reactive dyes by dried Rhizopus arrhizus: determination of column capacity. Journal of Hazardous Materials, 143 362-71.
[17] Mohamed Ali Wahab, Salah Jellali, Naceur Jedidi., (2010). Effect of temperature and pH on the biosorption of ammonium onto Posidonia oceanica fibers: Equilibrium, and kinetic modeling studies. Bioresource Technology, 101 8606–15.
[18] B. Ismail, S. T. Hussain, S. Akram., (2013). Adsorption of methylene blue onto spinel magnesium aluminate nanoparticles: Adsorption isotherms, kinetic and thermodynamic studies. Chemical Engineering Journal, 219 395-402.
[19] I. Langmuir., (1916). The constitution and fundamental properties of solids and liquids, 618 J. Am. Chem. Soc. 38, 2221–95.
[20] H.M.F. Freundlich., (1906). Over the adsorption in solution, Z. Phys. Chem. 57A 620 385–470.
[21] Woolard, J Strong, CR Erasmus., (2002). Evaluation of the use of modified coal ash as a potential sorbent for organic waste streams. Applied Geochemistry, 17 1159-64.
[22] Vadivelan V, Kumar KV., (2008). Equilibrium, kinetics, mechanism, and process design for the sorption of methylene blue onto rice husk. Journal of Colloid and Interface Science, 286, 90-100.
[23] S., Cengiz., L., Cavas., (2008). Removal of methylene blue by invasive marine seaweed: Caulerpa racemosa var. cylindracea. Bioresource Technology, 99 2357-63.
[24] A., Gurses, S., Karaca, C., Dogar, R., Bayrak, M., Acikyildiz, M., Yalcin., (2004). Determination of adsorptive properties of clay/water system: methylene blue sorption. J of Colloid and Interface Sci, 269 310- 314.
[25] N. Kannan, M.M. Sundaram., (2001). Kinetics and mechanism of removal of methylene blue by adsorption on various carbons—a comparative study, Dyes Pigments, 51 25–40.
[26] B. Bestani, N. Benderdouche, B. Benstaali, M. Belhakem, A. Addou., (2008). Methylene Blue and iodine adsorption onto an activated desert plant, Bioresour. Technol. 99, 8441–44.
[27] Largergren S., (1898). About the theory of so-called adsorption of soluble substances Vetenskapsakademiens. Handlingar, 24 1-39.
[28] Y.S. Ho, G. MacKay., (1999). Pseudo-second order model for sorption processes, Process, 644 Biochem, 34 451–65.

IOP Conf. Series: Materials Science and Engineering **323** (2018) 012007 doi:10.1088/1757-899X/323/1/012007

Synthesis, antioxidant and antibacterial activities of 3-nitrophenyl ferrocene

S Benabdesselam[1*], H Izza[2], T Lanez[3] and E K Guechi[4]

[1]Laboratoire de génie de l'eau et de l'environnement en milieu saharien, Faculty of Applied Sciences, University of Ouargla, Ouargla, Algeria.
[2]Laboraoitre de génie des procédés, Faculty of Applied Sciences, University of Ouargla, Ouargla, Algeria.
[3]Laboratory of Valorisation and technology of Saharan resources, University of El-Oued, El-oued, Algeria.
[4]Laboratory of environmental engineering, Faculty of Engineering , University of Annaba, Annaba, Algeria.

*Email: ben_soulef@yahoo.fr

Abstract. The current work aims in its first part to synthesize 3-nitrophenylferrocene after diazotizing nitroaniline in the meta position by the sodium nitrite and the formation of the corresponding diazonium salt: 3-nitrobenzendiazonium sulfate, then the salt in solution was added to the ferrocene for the purpose of introducing the nitrophenyl moiety thereon (arylation) and the formation of 3-nitrophenylferrocene. The second part is devoted to the study of the antioxidant activity of 3-NPF by applying the trapping test of superoxide radical using cyclic voltammetry, the free radical DPPH trapping test by spectrophotometry. The results showed that 3-nitrophenylferrocene has a scavenging effect of DPPH radical with IC50= 1.44mg/ml, superoxide radical with IC50=5.38mg/ml. The third part is devoted to the study of antibacterial activity of the synthesized compound tested on four strains of bacteria: Escherichia coli, Pseudomonas aeruginosa, Enterococcus faecalis and Klebsiella pneumoniae. The obtained results clearly showed that 3-nitrophenylferrocene has low activities on the four bacterial strains with diameters of inhibition zones do not exceeding 17 mm at concentrations of 25mg /ml.

1. Introduction

Ferrocene synthesised by kealy and pauson in 1951 [1] is an organometallic sandwich compound with iron in the formal oxidation state of +2, it has been object of continuous interest both in fundamental and applied research areas for many years. Recent advanced applications of Ferrocene have been also proposed in electronics, material sciences and medicinal chemistry [2]. Research has shown that ferrocene derivatives exhibit cytotoxic [3-4], anti-tumor [5, 6], antifungal [7], antibacterial [8] and antioxidant activities [9-11].

Oxidative stress is the cause of many diseases, including cancer, cardiovascular disease [12]. The body's defence system against cell damage induced by the high level of oxidative stress includes endogenous non-enzymatic natural antioxidants: glutathione and exogenous: ascorbic acid (vitamin C), tocopherol (vitamin E), and enzymatic antioxidants: superoxide dismutase, catalase and glutathione

IOP Conf. Series: Materials Science and Engineering **323** (2018) 012007 doi:10.1088/1757-899X/323/1/012007

reductase [13]. Researchers do not cease to develop new synthetic molecules that exhibit antioxidant activity to be used in many industries such as the food industry, cosmetics industry ...etc.
In this work, we synthesized 3- nitrophenylferrocene (3-NPF) and studied his antioxidant and antibacterial effect. The antioxidant activity of this compound is measured by an electrochemical method (Superoxide test) and chemical method (DPPH test), the antibacterial activity is tested on four strains of bacteria: Escherichia coli, Pseudomonas aeruginosa, Enterococcus faecalis and Klebsiella pneumoniae.

2. Materials and methods

2.1. Reagents
3-nitroaniline, Sodium nitrite, Magnesium sulfate, Ether, Petroleum ether, Sulphuric acid (98%). Methanol (99%), Dimethylformamide (99.9 %) were acquired from Biochem Chemopharma (Canada) , Dimeyhulsulfoxide (99.9 %) , Tetrafluoroborate tetrabutylammonium (99%) from Sigma-Aldrich, α -Tocopherol (97%) , 1,1- Diphenyl -2- picrylhydrazyl (DPPH) (99%) from Merck.

2.2. Synthesis
The overall synthesis strategy of 3-NPF is represented by the scheme illustrated in figure 1.
A mixture of 10.7g (0.077mol) of 3- nitroaniline, 16.7 ml of concentrated sulphuric acid and 22.90 ml of water was stirred for 1 hour then it was cooled to 0-5°C, diazotized with a solution of 5.5g (0.08 mol) of sodium nitrite in 12.25 ml of water. The cold solution was filtred and the filtrate was added with stirring to a solution of 13.08g (0.070mol) of ferrocene in 316.45 ml of ether. The mixture was stirred for 3 h at room temperature. The two layers were separated. The aqueous layer was extracted with Et_2O, and then the organic layer washed with water, dried with $MgSO_4$. Concentration in vacuo and purification by column chromatography gave 3-NPF as orange plates with Y= 56%. (R_f = 0, 73).

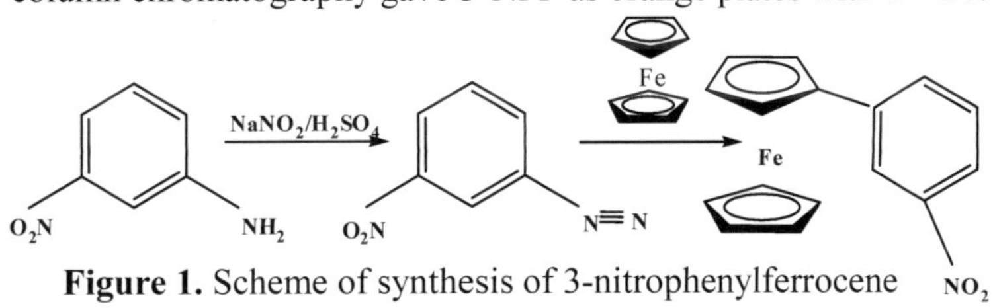

Figure 1. Scheme of synthesis of 3-nitrophenylferrocene

3. Antioxidant activity

3.1. Scavenging activity of free radical DPPH
The inhibitory activity of DPPH radical is evaluated using the stable radical DPPH and following the method described by Brand- Williams and al [14] with some modifications. 3 - NPF in 0.1ml of DMSO was added to 1.9 ml of DPPH solution in methanol. The samples are incubated for 30 minutes and the absorbance of DPPH solutions was measured at 517nm.Results are expressed and calculated using the equation (1):

$$I\,(\%) = \frac{A_0 - A_s}{A_0} \times 100 \quad (1)$$

I (%): inhibition capacity of DPPH, A_0: absorbance of the DPPH solution without sample
A_s: absorbance of the test sample mixed with DPPH solution.

3.2. Scavenging activity of superoxide anion radical ($O_2^{\bullet -}$)
The inhibitory activity of the superoxide radical anion is determined by cyclic voltammetry, this test is based on the method of Bourvellec and al with some modifications [15].

Voltammetric measurements are carried out using 40 VoltaLab PGZ301 model (Radiometer Analytical SAS), a three-electrode system in an electrochemical cell of volume V = 25 ml. A saturated calomel electrode (SCE) is used as a reference electrode, a platinum wire as an auxiliary electrode and a glassy carbon electrode about 0.013cm^2 section as the working electrode. The data acquisition is performed by a computer Pentium IV (3.0GHz CPU and 1 GB RAM) using VoltaMaster 4 software, Version 7.08 (SAS analytical radiometer).

Before use, the working electrode was polished, washed with distilled water and dried with absorbent paper. This cleaning procedure is always applied before each electrochemical measurement. The superoxide anion radical is generated by the commercial molecular oxygen dissolved in DMF containing 0.02M of Bu_4NBF_4 at room temperature (28 ± 1 ° C). The scanning rate is maintained at 100 mV / s. The range of applied potential was -1.6V to 0.0V vs SCE.

3-NPF is added to the superoxide radical dissolved in DMF and voltammograms were recorded (figure 4).

The studied compound's ability to scavenge superoxide radicals ($O_2^{\bullet -}$) is calculated using the following equation (2):

$$I\,(\%) = \frac{Ip_0 - Ip_s}{Ip_0} \times 100 \qquad (2)$$

wherein: Ips and Ip0 are the intensity of the anodic peak current of the oxygen respectively without and with test compound.

I (%): Inhibitory capacity of superoxide.

4. Antimicrobial activity

4.1. Microorganisms

Four strains of bacteria were used in this study including Escherichia coli ATCC25922, Pseudomonas aeruginosa, Klebsiella pnenumoniae ATCC700603, Enterococcus faecalis ATCC29212.

4.2. Agar well diffusion assay

The qualitative antibacterial assay of 3-NPF was carried out by the disc diffusion method [16], all the bacterial strains mentioned above were incubated at 37°C for 24 h by inoculation into Mueller-Hinton (M-H) broth. The samples were prepared by dissolving 3-NPF in DMSO to obtain solutions with concentrations: 0.2 mg/ml, 2, 10, 20,25mg/ml. A bacterium was spread on the Muller-Hinton Agar (MHA) medium in a petri dish, which was incubated at 37°C for 24h to obtain uniform colonies of the bacterium. Then 4-5 morphologically identical colonies of the bacterium were put into a test tube containing autoclaved normal saline solution (5ml).Then the bacterial suspension was transferred into petri disches that contained autoclaved MHA medium. We file on the surface of the Petri dishes filter paper disks of diameter 5 mm impregnated with the 3 - NPF solutions prepared at various concentrations. After incubing the petri dishes at 37°C for 24h, inhibition zones (ZI) were measured. Each test was repeated at least for three times and the mean was calculated.

5. Results and discussion

5.1. Synthesis

The synthesis of 3-nitrophenylferrocene takes place in two stages, the first stage being a diazotization of 3-nitroaniline by sodium nitrite and the second step is an arylation of ferrocene whose objective is the introduction of nitrophenyl group on the latter by diazonium salt intermediate; nitrobenzenediazonium sulphate formed in the first step.

This compound was synthesized according to the following procedure, 10.7 g of 3-nitroaniline dissolved in 22.90 ml of water + 16.7 ml of H_2SO_4 and diazotized with 5.5 g of sodium nitrite. 13.08 g of ferrocene in 316.45 ml of ether was added to the filtrate (3-nitrobenzenediazonium sulphate). After 3 hours of stirring under the room temperature, the organic phase is recovered, washed with water, dried and evaporated to obtain the reaction product. Column chromatography and recrystallization in methanol allows recovering the products as red orange solid.

Its proton NMR spectrum shows different peaks, a peak at 4.05 ppm corresponding to 5H of the pentadienyl (C_5H_5) ferrocene ring, a peak at 4.40ppm corresponds to 2H linked to the C2 or C3 atom, a peak at 4.71ppm corresponding to 2H linked to the C3 atom or the C2 atom, a triplet at 7.43 ppm corresponds to a proton linked to the C9 atom with a coupling coefficient j equal to 7.98Hz, a doublet at 7.74ppm corresponds to a proton linked to the C10 atom with a coupling coefficient equal to 7.70Hz, a doublet at 8.00ppm corresponds to a proton bound to the C8 atom with a coupling coefficient equal to 8.34Hz and a singlet peak corresponds to a proton linked to the C8 atom, atom C6.

By combining the information extracted from the 13 C and DEPT NMR spectra with those obtained from the proton spectrum of the compound, we can deduce that the compound does not contain CH_2, and CH_3, it's contains 7 (CH) , 3 (C) and 13H.

5.2. *Antioxidant study*

5.2.1. *Scavenging activity of free radical DPPH*

Free radical scavenging activity of 3-NPF was determined through DPPH assay. DPPH (1,1-diphenyl-2-picrylhydrazyl) is a stable free radical. It accepts an electron or hydrogen to form a diamagnetic species, and as a result, its purple colour (λmax 517nm) changes into pale yellow. When a sample containing a substance with radical scavenging potential is mixed with DPPH, reaction takes place and concentration of DPPH decreases causing a decrease in absorbance at 517nm. Therefore, a stronger free radical scavenger causes a greater decrease in the absorbance [17]. The concentration of antioxidant required to reduce 50% of the initial concentration of oxidizing agent (DPPH) expressed by IC50 is a widely used parameter for measuring the antioxidant effect. The IC50 values were calculated graphically by plotting percentage inhibition against concentration of test compound. A low IC50 means high antioxidant activity. All measurements were performed in triplicate to ensure reproducibility of the results. The results obtained are shown in table 1.

5.2.2. *Scavenging activity of superoxide anion radical ($O_2^{\bullet -}$)*

The superoxide anion is a weak oxidant, but it is capable to produce powerful and damaging hydroxyl radicals and singlet oxygen, both of which contribute to oxidative stress. In this work, we chose the superoxide due to its presence in the human body as well as its long half-life. The superoxide anion radical was generated by one electron reduction of the commercial molecular oxygen (O_2) dissolved in DMF at room temperature (28±1°C). The resultant CV response of the medium (DMF+0.02 M Bu4NBF4) and the superoxide radical in the same medium are shown respectively in figure 2 and figure 3.

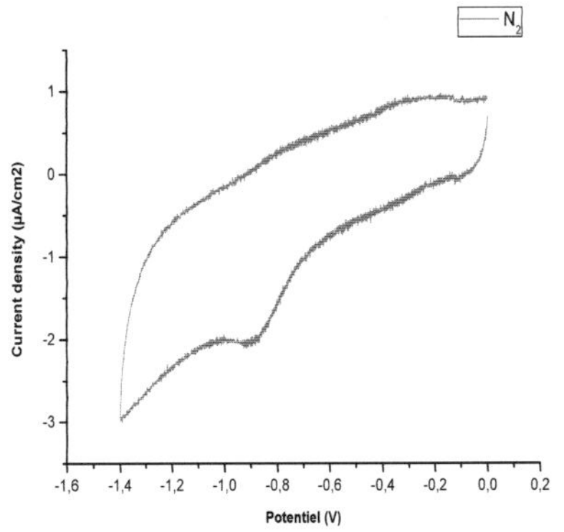

Figure 2. Cyclic voltammogram of the medium (DMF+0.02 M Bu4NBF4).

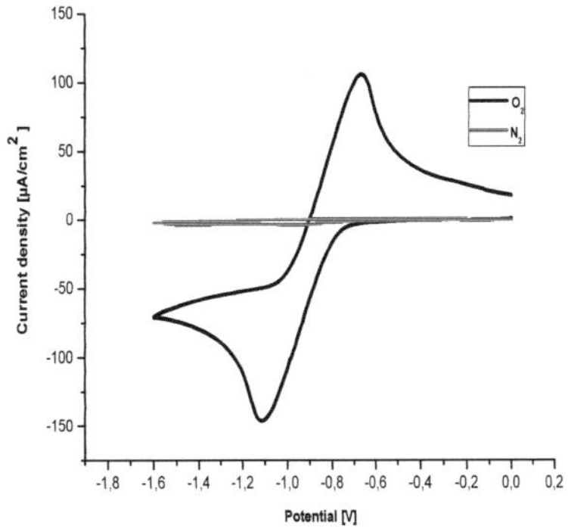

Figure 3. cyclic voltammogram of $O_2^{\bullet -}$ in DMF + 0.02 M Bu4NBF4, on CV as working electrode at 28° C ± 1 C with scan rate of 100 mV/s.

- *α-Tocopherol*

The figure 4 illustrates the cyclic voltammogrames of $O_2^{\bullet -}$ in the presence of different concentrations of α-Tocopherol in DMF.

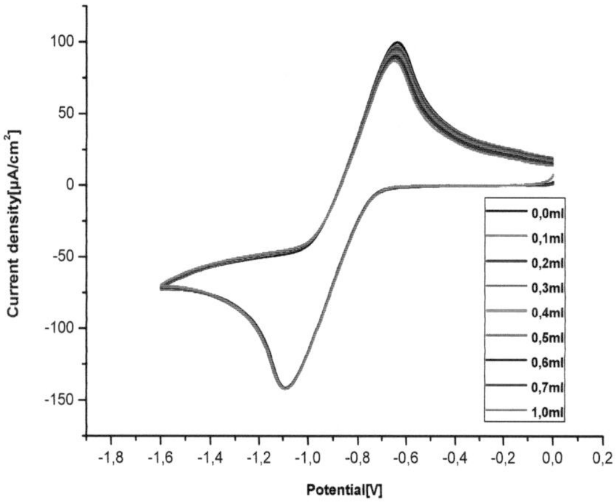

Figure 4. Voltammogrames of $O_2^{\bullet -}$ in the presence of different concentrations of α -Tocophérol in DMF + 0.02 M Bu_4NBF_4 on CV as the working electrode at 28 ° C ± 1 with a scan rate of 100mV / s.

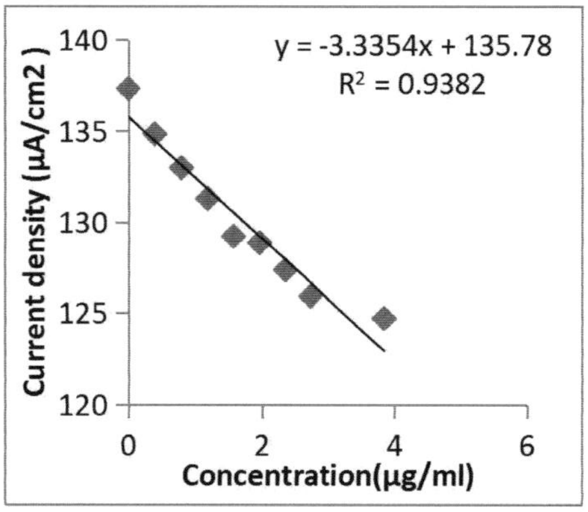

Figure 5. Change in current intensity as a function of the concentration of α-Tocopherol.

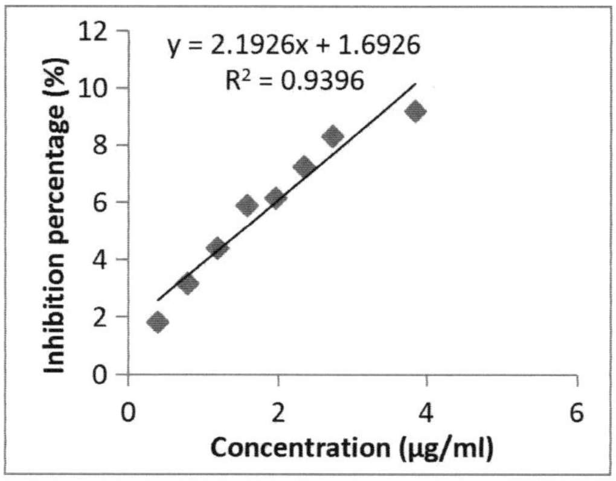

Figure 6. percent inhibition depending on the concentration of α-Tocopherol.

Figures 5and 6 illustrate the change in current intensity as a function of the concentration of α-Tocopherol and the percent inhibition evolution depending on the concentration of α-Tocopherol.

- *The studied compound: 3-nitrophenylferrocène*

The figure 7, 8 and 9 present respectively the cyclic voltammogrames of $O_2^{\bullet -}$ in the presence of different concentrations of 3-NPF in DMF , the change in current intensity as a function of the concentration of 3- NPF and the percent inhibition evolution depending on the concentration of 3-NPF.

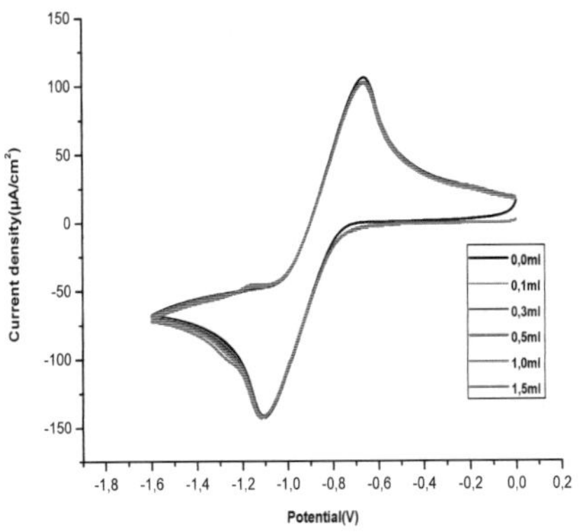

Figure7. Voltammograms of $O_2^{\bullet -}$ in the presence of different concentrations of 3- NPF in DMF + 0.02 M Bu_4NBF_4 on CV as the working electrode at 28 °C ± 1 with a scan rate of 100mV / s.

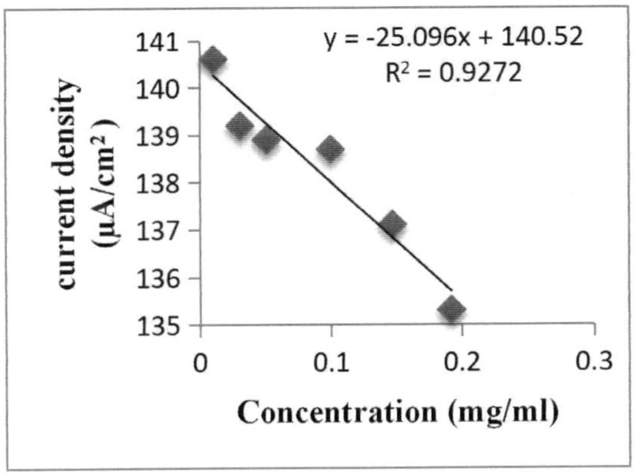

Figure 8. Change in current intensity as a function of the concentration of 3- NPF.

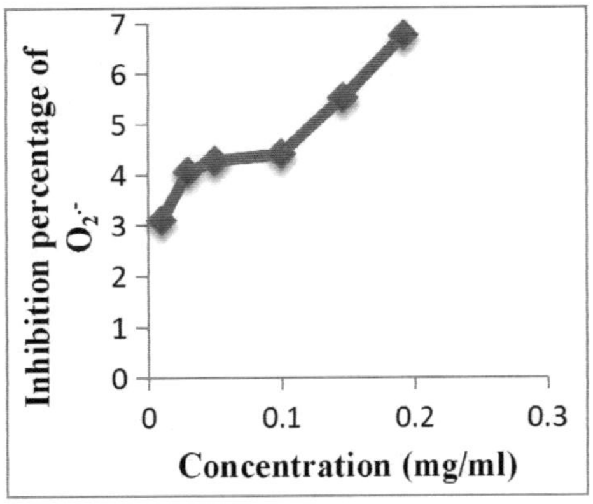

Figure 9. percent inhibition evolution depending on the concentration of 3-NPF.

The decrease of the anodic peak current of $O_2^{\bullet -}$ suggests that the ferrocene derivative reacts irreversibly with $O_2^{\bullet -}$

For each antioxidant, a series of values is determined from the CVs recorded for increasing antioxidant concentrations (table1).

From the table 1, It can be seen that the IC50 values of 3-NPF showed an antioxidant capacity with an IC50 of 5.38mg/ml for the superoxide radical and an IC50 of 1.44 for the DPPH radical.

Table 1. IC50 values obtained by the DPPH test and superoxide test.

Compound	Methods	Equation	R^2	IC50(mg/ml)
3-NPF	DPPH	Y=36.30x-2.481	0.983	1.44
	$O_2^{\bullet -}$	Y=17.293x+3.152	0.994	5.38
α-tocopherol	DPPH	Y=1.56x+29.093	0.995	0.0134
	$O_2^{\bullet -}$	Y=2.955x+2.269	0.961	0.016

5.3. Antibacterial study

Antibacterial activity of 3-NPF was determined against four strains of micro-organisms. Mean zones of inhibition were calculated and the results are displayed in figure 10.

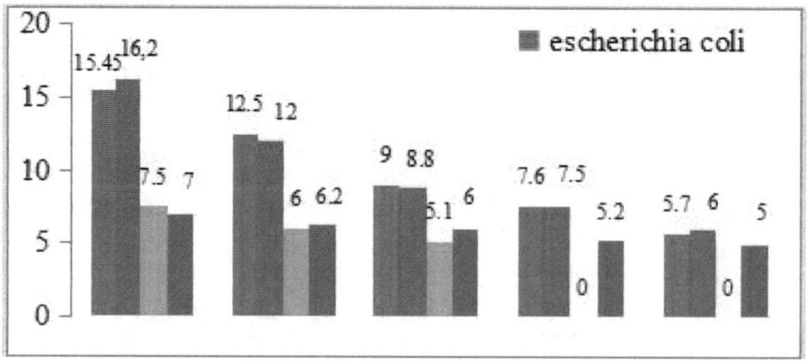

Figure 10. Mean inhibition zones of the bacteria at different concentrations of 3-NPF.

We notice from the results that the four bacterial strains are insensitive towards 3 - nitrophenylferrocene, 3 - NPF has an important activity on Escherichia coli with inhibition zone diameter of 15.45 mm for 25mg / ml and 5.7 mm to 0.2 mg / ml. For Pseudomonas aeruginosa, 3 - NPF has antibacterial activity against it close to that obtained with Escherichia coli, and it results in the same inhibition zone diameters for all concentrations. 3 - NPF has negligible activity on Enterococcus faecalis and Klebsiella pneumoniae.

6. Statistical analysis

The results were presented as the mean values ± SD. Correlation analyses between the antioxidant activities were carried out using the correlation and regression program in the EXCEL program.

7. Conclusion

In conclusion, the antioxidant activity of 3- nitrophenylferrocene was assessed by using electrochemical and chemical methods. 3 - nitrophenyferrocène has an IC50 equal to 1.44 mg / ml evaluated by the DPPH test and IC50 equal to 5.38 mg / ml measured by the superoxide scavenging test. Our results indicate that 3 - NPF has a low activity to trap the radical 2, 2- diphenyl- 1 - picrylhydrazyl and superoxide. The results of the antibacterial activity showed that 3-nitrophenylferrocene has low activities on the four bacterial strains with diameters of inhibition zones do not exceeding 17 mm at concentrations of 25mg / ml.

References

[1] T.J. Kealy, P.L. Pauson , (1951), A new type of organo iron compound,Nature 168 1039–40.

[2] J. Vera, G. L. Ming, A. Santana, J. Matta, E. Mel´endez, (2011) Vectorized ferrocenes with estrogens and vitamin D2: synthesis, cytotoxic activity and docking studies,The Royal Society of Chemistry 40 9557–65

[3] I R. Acovino, J.V. Caso, F. Rapuano, A. Russo, M. Isidori., M. Lavorgna, G. Malgieri , (2012) Physicochemical Characterization and Cytotoxic Activity Evaluation of Hydroxymethylfer,rocene:β Cyclodextrin Inclusion Complex, Molecules 17 6056-70

[4] B. Imre Károlyi, S. Bősze, E. Orbán , P. Sohár., L. Drahos , E . Gál , A. Csámpai, (2012) Acylated mono-, bis- and tris- Cinchona-Based Amines Containing Ferrocene or Organic Residues: Synthesis, Structure and in Vitro Antitumor Activity on Selected Human Cancer Cell Lines Molecules 17 2316-29

[5] B. Zhou, J. Li,_ BJ. Feng, Y. Ouyang, YN. Liu, F. Zhou; (2012)Journal of Inorganic Biochemistry 116 19–25

[6] A.S. Altaf, A. Badshah, ud-Din I., Bhajan Al., kamal S.et Ullah S. ; (2013), DNA interaction, Antibacterial and Antifungal studies of 3-nitrophenylferrocene ,J. Chem. Soc. Pak 35 922-28

[7] H. R. Azadar, A. Badshash, M. Sohail, B. Lal , A. Ali Altaf , (2013), Synthesis, chemical characterization, DNA interaction and antioxidant studies of ortho, meta and para fluoro substituted ferrocene incorporated selenoureas Inorganica Chimica Acta 402 133–39

[8] G.L. Xi, Z.Q. Liu, (2014) Solvent-free Povarov reaction for synthesizing ferrocenyl quinolines: Antioxidant abilities deriving from ferrocene moiety, European Journal of Medicinal Chemistry 86 759-68
[9] G. Nabi, Z.Q. Liu, (2011) Radical-scavenging properties of ferrocenyl chalcones, Bioorganic & Medicinal Chemistry Letters 21 944–46
[10] J. Zhang, R. Liu, (2011) J.Chem.Soc.Pak 33 922- 28
[11] R. Kohen, A. Nyska, (2002) Oxidation of Biological Systems: Oxidative Stress Phenomena,Antioxidants, Redox Reactions, and Methods for Their Quantication ,Toxicologic Pathology, 30 620–50
[12] J. N. Temple, (2000) Antioxidants and disease: More questions than answers, Nutrition Research 20 449-59
[13] B. Halliwell, (1996) Antioxidants in human health and disease, Anna Rev. Nutr 16 33-50
[14] W. Brand-Williams, M.E. Cuvelier, C. Berset, (1995) Use of a Free Radical Method to Evaluate Antioxidant Activity, Science and Technology 28 25–30
[15] C. Le Bourvellec, D. Hauchard, A. Darchen, J. L. Burgo, M. Abasq; (2008) Validation of a new method using the reactivity of electrogenerated superoxide radical in the antioxidant capacity determination of flavonoids, Talanta 75 1098–03
[16] AA.Marmonier, (1987), Bactériologie médicales, Techniques usuelles. Chap 4. Antibiotiques. Techniques de diffusion en gélose méthode des disques © SIMEP SA-PARIS, France 238-44
[17] J. Bokhari, M.R. Khan, M. Shabbir , U. Rashid , S. Jan, J.A. Zai , (2013) Spectrochimica Acta Part A: Molecular and Biomolecular Spectroscopy 102 24–29

IOP Conf. Series: Materials Science and Engineering **323** (2018) 012008 doi:10.1088/1757-899X/323/1/012008

Effect of surfactant on single drop mass transfer for extraction of aromatics from lubricating oils

H. IZZA[1*], S. BEN ABDESSALAM[2] and M. KORICHI[3]

[1]Univ Ouargla, Fac. des Sciences Appliquées, Laboratoire de Génie des Procédés, Algérie
[2]Univ Ouargla, Fac. des Sciences Appliquées, Laboratoire de Génie de l'eau et de L'environnement en milieu saharien, Algérie
[3]Univ Ouargla, Fac. des Sciences Appliquées, Dept. de Génie des Procédés, Lab. Dynamique, Interactions et Réactivité des Systèmes, BP 511 Ouargla, Algérie

*E-mail address: hidaya_iza@yahoo.fr

Abstract. Solvent extraction is an effective method for the reduction of the content of aromatic of lubricating oil. Frequently, with phenol, furfural, the NMP (out of N-methyl pyrrolidone). The power solvent and the selectivity can be still to increase while using surfactant as additive which facilitates the separation of phase and increases the yeild in raffinat. Liquid-liquid mass transfer coefficients for single freely rising drops in the presence of surfactant in an extraction column have been investigated. The surfactant used in this study was sodium lauryl ether sulfate (SLES). The experiments were performed by bubbling a solvent as a series of individual drops from the top of the column containing furfural-SLES solution. The column used in this experiment was made from glass with 17 mm inner diameter and a capacity of 125ml. The effects of the concentration of surfactant on the overall coefficient of mass transfer was investigated.

1. Introduction

The process of manufacturing lubricating oil base stocks requires the removal of aromatic hydrocarbons from vacuum distillates [1,2].Under atmospheric conditions , the crude oil Residue is transferred firstly to a vacuum distillation column, and then separated into different lube oil cuts which are characterized by their boiling range and viscosity.

N – methyl –2 – pyrrolidone, furfural, phenol and liquid sulfur dioxide are the largest used solvents [3-5].

Liquid- liquid extraction is an important separation process. It is a widely used separation process in industry.

Some studies on the use of surfactants as additives in liquid-liquid extraction have been mentioned in literature [6-9]. A surfactant is a substance that significantly decreases the surface tension of a liquid when added in small quantities. If a small amount of surfactant is added during the extraction process, both masses of the transfer coefficient and the interfacial area may be affected.

Several investigators [10-12] suggest that the effects of surfactants cause a decrease in mass transfer rate. The results indicate that surfactants often reduce the mass transfer coefficient but increase the mass transfer area per unit volume.

Abolghasemi and al [13] in their studies performed on extraction, the decrease in the mass transfer coefficient by surfactant was attributed either to the decrease in interfacial mobility and internal circulation or to the blocking action of surfactant molecules at the interface it has been generally assumed that such activity should facilitate mass transfer across the interface.

Audeh and al [14], discovered an improved furfural extraction process for lube oil base stock containing aromatic type material by the addition of demulsifiers to furfural to facilitate phase separation, selectivity, and increase the raffinate yield while maintaining the same raffinate measured by raffinate refractive index.

They suggested a wide variety of surfactant. Among these are included those of the type having the formula $R\text{-}CONH\text{-}CH_2CH_2SO_3Na$.

For a class of the sulfonic acid salt surfactant, the metal may be taken from Groups IA, HA and HE of the Periodic table, as, for example, sodium, calcium, barium and zinc, wherein the alkyl contains 1 to 30 carbon atoms, preferably 10 to 20 carbon atoms.

We present in this paper our investigations regarding the use of sodium lauryl ether sulfate (SLES) as surfactant.

Our choice for this particular class of surfactants (ethoxylated anionic surfactants) was also motivated by the fact that in spite of their wide range of applications, their self-aggregation in aqueous solutions is not clearly understood. For instance, it may be anticipated that the addition of hydrophilic oxyethylene (OE) groups adjacent to the anionic sulfate moiety would increase the overall hydrophilic character of the molecule and increase the CMC accordingly [15].

2. Experimental procedures

2.1. Chemicals

The SPD (spindle distillat) and furfural were provided by a refinery of Arzew (Algeria). The feedstock for vacuum distillation unit was atmospheric residue produced from light Hassi Messaoud crude oil. Their physical properties are listed in table 1.

Anhydrous calcium chloride and Cyclohexane were purchased from Fluka.

(SLES) $C_{12}H_{25}(OC_2H_4)_2SO_4Na$ with trade name Galaxy LES 70 with 70 % active matter was provided by Galaxy Surfactants Ltd. Surfactant solutions were prepared with furfural solvent.

2.2. Experimental section

The SPD and furfural were provided from Arzew refinery (Oran-Algeria). The SPD sample was collected from the light crude oil of Hassi Messaoud.

The extraction is carried out in a column of extraction out of glass, of a capacity of 125ml with a diameter of 17mm, the temperature is regulated, by the circulation of oil and the solvent (furfural) in a bath thermostat, the thermostat will have to keep the temperature of constant measurement to 95 C°.

The counter-current is carried out by gravity, it is to say by difference in density between the phases. The phase continues is pumped in bottom of the column, and the dispersed phase is pumped at the top of the column, the time of flow of the droplet which represents the dispersed phase is measured by chronometer.

The decantation is carried out during one hour of time in order to reach balance. After that one separates the sub-base (phase of extract) in Matras and the road base (phase D in the column from extraction.

The raffinate obtained is washed with distilled water heated at a temperature of 90°C in order to remove from furfural. The operation is repeated six times. The washed product is dehydrated by the calcium dichloride.

The chemical system used in these experiments is lubricating oil/aromatics/furfural/SLES, where the lubricating oil is considered as the continuous phase (feed phase), furfural as the dispersed phase, aromatics as a solute and SLES as a surface active agent (surfactant). At the beginning of the operation, lubricating oil and furfural are brought to saturation. SLES with known and different concentrations are added to the furfural phase (solvent phase). The mass transfer direction in this system is from raffinate phase to extract phase, the Schematic diagram of experimental setup is shown in figure 1.

The Experimental conditions are:
Solvent ratio relative to oil equal to 1
Extraction temperature: 95 ° C
Surfactant concentration in the solution of furfural: 0; 0.01%; 0.05%; 0.1%

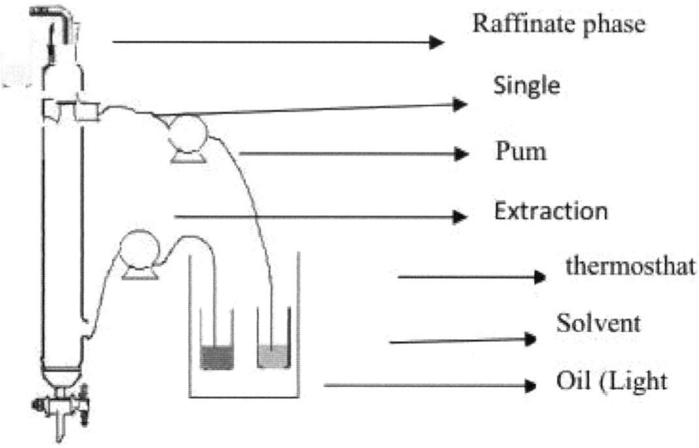

Figure 1. Schematic diagram of experimental setup.

The column is provided with a device of dispersion to improve contacts it with diameter of 3mm as shown in figure 2.

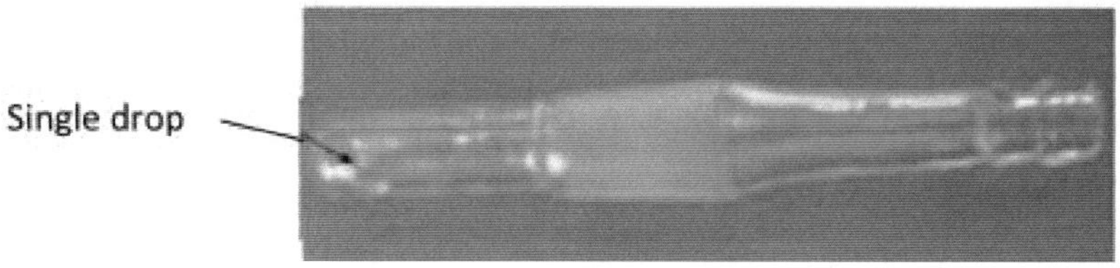

Figure 2. Dispersing device.

2.3. Determination of the composition according to the cycle number of aromatic
To determine, the selectivity, the oil composition in aromatic hydrocarbons (mono, di, poly), we used the method of Burdett [16]. This method takes into account the fact that among the hydrocarbons, only aromatics give an absorption in ultraviolet, with a maximum ratio, which shifts to longer wavelengths when the condensed nucleus number increases:

Mono aromatic : 197 nm
Diaromatic : 230 nm
Tri et polyaromatic : 260 nm

The compositions of aromatic (mono, di, Poly) were measured by absorption spectroscopy with ultraviolet using ASTM D 2008 standard test method [17].

The absorption coefficients a_1, a_2 and a_3 were measured at three wavelengths with the following characteristics (concentration of 1 g/l for a cell thickness of 1 cm). The raffinate and extract phases were diluted with cyclohesane. In current study Spectrophotometer (spectroScan 80DV-UV/Vis) was used.

The mass percentages of aromatics (Mono, Di, Poly) are determined according to the following equation:

$$X_{A1} = 10^{-4}\left(20.17a_1 - 2.59a_2 - 5.47a_3\right) \quad (1)$$

$$X_{A2} = 10^{-4}(-0.68a_1 + 12.83a_2 - 3.00a_3) \quad (2)$$

$$X_{A3} = 10^{-4}(-0.05a_1 - 0.45a_2 + 17.98a_3) \quad (3)$$

The saturated compound was calculated using following equation:

$$X_B = 100 - X_T \quad (4)$$

2.4. Mass transfert calculation model

In this process, the surfactant is transfered from drops to the solvent as dispersed phase. From the mass balance of SLES in single drop, the following equation is obtained:

$$K_d a_d (C_d - C^*_d) = -(V_d C_d) \quad (5)$$

were k_d is the overal mass transfer coefficient for the dispered phase (fufural), and a_d and V_d are surface area and volume of a drop, C_d is the solute concentration in the dispersed phase and C^*_d is the solute concentration in the dispersed phase in equilibrium with its concentration in the continuous phase. If the solute concentration in the continuous phase is symbolized by $C_{c'}$, then C^*_d and C_c can be correlated by the following expression:

$$C^*_d = mC_c \quad (6)$$

where m is the equilibrium distribution coefficient.

$$K_d = \frac{V_d}{a_d t_d} \ln\left(\frac{C_{d1}}{C_{d2}}\right) \quad (7)$$

with the assumption that the drop is spherical, its surface area and volume are

$$a_d = \pi d_d^2 \quad (8)$$

$$V_d = \left(\frac{\pi}{6}\right) d_d^3 \quad (9)$$

Substituting Eqs. (8) and (9) into Eq. (7) gives:

$$k_d = \left(\frac{d_d}{6\,t_d}\right) \ln\left(\frac{C_1}{C_2}\right) \quad (10)$$

In this study, the drop diameter (d_d) was calculated from the following equation (McCab, Smith, and Harriot 1993:

$$d_d = \left[\frac{6 d_0 \sigma g_C}{g(\rho_d - \rho_c)}\right]^{1/3} \quad (11)$$

The drop rise time (t_d) in Eq. (6) was experimentally measured by using a stopwatsh. C_{d1} and C_{d2} are the measured solute concentrations in the feed and raffinate solutions, respectively.

3. Results and discussions

Table 1. Values of yield; compositions in mono-aromatics (**X_{A1}**), di-aromatic (**X_{A2}**), poly-aromatics (**X_{A3}**) and saturates (**X_{BR}**) in the raffinate phase.

Fraction	X_{A1} (wt.%)	X_{A2}(wt.%)	X_{A3}(wt.%)	X_{AT}(wt.%)	X_{BR}(wt.%)
SPD	14.60	5.16	3.48	23.24	76.77
R1	9.68	2.62	1.73	14.03	85.98
R2	10.04	3.18	1.95	15.17	84.84
R3	9.032	2.23	1.38	12.65	87.36
R4	9.54	2.43	1.47	13.43	86.57

IOP Conf. Series: Materials Science and Engineering **323** (2018) 012008 doi:10.1088/1757-899X/323/1/012008

In view of the results obtained, it is noted that:

A different surfactant concentration, polyaromatic and diaromatic extracts are better than monoaromatic. It was found that the maximum polyaromatic extraction was carried out using SLES concentration at 0.05%wt.

The diaromatic extraction was for a SLES concentration of 0.05% better quality in raffinate was obtained with the SLES of 0.05%, whose total aromatic is of 12.64%. From which it is deduced, the decrease in the aromatic content that is due to the increase in the selectivity of furfural for SLES concentration of 0.05%.

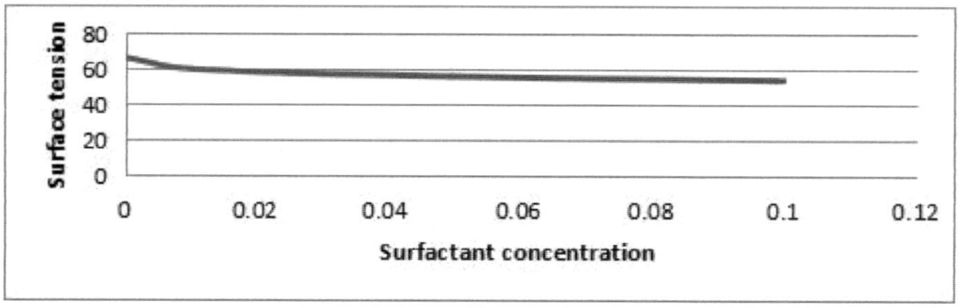

Figure 3. Variation of the surface tension as a function of the surfactant concentration.

From figure 3, it can be seen that the surface tension decreases with increasing surfactant concentration. This decrease is slower in the area of high concentrations.

The surfactant is capable of decreasing the surface tension due to adsorption and orientation of the molecules on the phase separation surface.

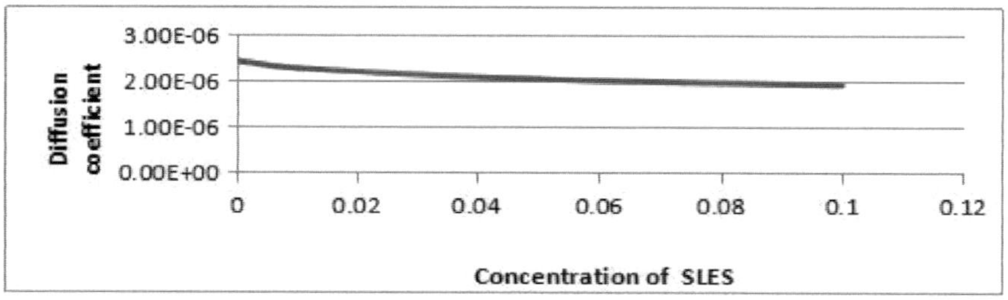

Figure 4. Variation of the diffusion coefficient as a function of the concentration of SLES.

From figure 4, the addition of surfactant in the dispersed phase lowers the overall mass transfer coefficient (the diffusion coefficient).

Increasing the surfactant concentration decreases the diffusion coefficient value.

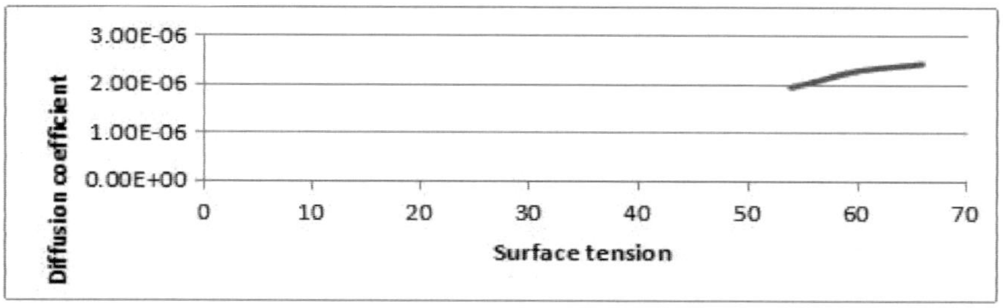

Figure 5. Correlation between diffusion coefficient and surface tension.

Figure 5 shows the correlation between the diffusion coefficient and the surface tension. The K_d value increases with increasing the surface tension value of the dispersed phase. The increase in the surface tension of the dispersed phase caused a high gradient of the surface tension between the continuous phase and the dispersed phase. This difference increases the interfacial turbulence between the continuous and dispersed phase, which increasing the value of the mass transfer coefficient.

IOP Conf. Series: Materials Science and Engineering **323** (2018) 012008 doi:10.1088/1757-899X/323/1/012008

4. Conclusion

The aim of the work presented is to extract the aromatic compounds from a petroleum cut with and without the presence of the surfactant.

The extraction of the aromatics is carried out in an apparatus making it possible to carry out the following two operations:

- An intimate contact between the two liquids: the solvent and the oily charge.
- A separation of the two phases obtained: the extract and the raffinate.

This differential contact between a dispersed phase and a continuous phase, which takes place by countercurrent flow with external energy supply, establishes a concentration gradient giving rise to a transfer of material, by continuous and progressive diffusion, over the whole Length of countercurrent.

The experiments carried out clearly show that, with the addition of different concentrations of surfactant in the dispersed phase, it is possible to:

- Ensure good phase separation.
- Reduction of the overall mass transfer coefficient (the diffusion coefficient), and consequently an improvement in raffinate quality.

References

[1] Coto B , vanGrieken R, Pena JL, Espada JJ, 2006 *Amodel to predict physical properties for light lubricating oils and its application to the extraction process by furfural*, Chem.Eng.Sci.61, pp 4381–92.

[2] Espada J.J, Coto B, vanGrieken R,Moreno J.M, 2008 *Simulation of pilot-plant extraction experiments to reduce the aromatic content from lubricating oil*, Chem. Eng.Process. 47, pp, 1398–1403.

[3] Fakhr-Hoseini SM, Tavakoli T, Hatamipour MS, 2009 *Extraction of aromatic hydrocarbons from lube oil using n-hexane as a co-solvent.* Sep Purif Technol; 66, pp, 167–70.

[4] Nelson, W.L 1978 *Petroleum Refining Engineering*. McGraw-Hill, New York

[5] Sequeira, A. Jr.: Lubricant Base Oil and Wax Processing. pp. 81–82. Marcel Dekker, New York (1994)

[6] Mulyono P, Muliati S 2007 *Effect of surfactant on Single drop Mass Transfer in liquid –liquid extraction*. AJChE, vol. **7**, N° 1, pp, 1–7

[7] Handlos, A.E, and Baron, T 1957 *Mass and heat transfer from drops in liquid-liquid extraction*, A.L.Ch.E. Journal,3, pp, 127–36.

[8] Lee Y.L, Maa.J.R, and Yang Y.M 1998 *The effect of a surfactant on mass transfer in extraction column*; J.Chem. Eng. Japan, 31, pp, 340–46.

[9] West F.B, Harrman, A.J, Chong, A.T, and Thomas, L.E.K (1952). *Addition agents and interfacial barriers in liquid-liquid extraction*, Ind. Eng. Chem., 44, pp, 625–31.

[10] Garner, F.H.,and Hale; A.R.(1953). «The effects of surface active agents in liquid extraction processes», Chem.Eng. Sci., 2, pp, 157–63.

[11] Huang, W.S., and kintner, R.C. (1969).« Effects of surfactants on mass transfer inside drops», A.I.CH.E. Journal, 15 pp, 735–74.

[12] Y.L.Lee, J.R.Maa, Y.M. Yang, the effect of surfactant on mass transfer in extraction colomn, J.Chem. Eng. Japan, 31 pp, 340–46 (1998).

[13] Abolghasemi, Hossein; Moosavian, Mohammad Ali; Radpour, Saeid Reza 2006 *The effects of a surfactant concentration on the mass transfer in a mixer-settler extractor*, Iran. J. Chem. Chem. Eng. Vol. **25**, No.4.

[14] Audeh, C.A; Israel, J; Heilwel, J.R1983: *White. Solvent extraction production of lube oil fractions* , US patent, 4381234

[15] Francesc Comelles, Isabel Ribosa, Juan José Gonziez, M. Teresa Garcia , 2014 *Micellization of sodium laurylethoxysulfate (SLES) and short chain imidazolium ionic liquids in aqueous solution*, Journal of Colloid and Interface Science 425, pp, 44–51.

[16] Burdett R.A 1954 *Determination of aromatic hydrocarbons in lubricating oil fractions by far-*

ultraviolet absorption spectroscopy. In Molecular spectroscopy, Institute of petroleum George Sell, Londres, pp, 30–41.
[17] ASTM Annual Book of Standard, Americain Society for Testing and Materials 2003 vol. **5**,West Conshohocken, Pennsylvania.

Nomenclature

Mw	Mean molecular weight
SG	Density (20°C/20°C)
RI	Refractive index(20°C)
m_A	Distribution coefficient of aromatic
m_B	Distribution coefficient of saturated
a_1	Absorption coefficient at longer wavelengths of 197nm
a_2	Absorption coefficient coefficient at longer wavelengths of 230nm
a_3	Absorption coefficient coefficient at longer wavelengths of 260nm
A_1	Monoaromatic hydrocarbon
A_2	Diaromatic hydrocarbon
A_3	polyaromatic hydrocarbon
A_T	Total aromatic hydrocarbon
B	Saturated hydrocarbon
X_{A1}	Monoaromatic concentration (%mas.)
X_{A2}	Diaromatic concentration (%mas.)
X_{A3}	Polyaromatic concentration (%mas.)
X_{AR}	Aromatic concentration in raffinat phase (%mas.)
X_{BR}	Saturated concentration in extract phase (%mas.)
X_{ER}	Saturated concentration in raffinat phase (%mas.)
a_d	Specific surface area of a drop
V_d	Volume of a drop
C_d	The solute concentration in the dispersed phase.
C^*_d	The solute concentration in the dispersed phase at equilibrium with its concentration in the continuous phase.
m	Coefficient of equilibrium distribution.
d_0	Diameter of the orifice for the dispersed phase
ρ_c	Continuous phase density (kg/m^3)
ρ_d	Dispersed phase density (kg / m3)
σ	Surface tension (mN / m)
g	Constant of gravity (m / s2)
d_d	Drop diameter of the dispersed phase (mm)
C_1	Initial aromatic charge concentration (spindle) (mol / m3)
C_2	Concentration of aromatics after extraction (mol / m3)
t_d	Droplet flow time of the dispersed phase (s)
Subscripts	
SPD	Spindle (light lubricating oil)
SLES	Sodium lauryl ether sulfate
R	Raffinate phase
E	Extract phase
Greek letters	
B	Selectivity
γ	Activity coefficient

IOP Conf. Series: Materials Science and Engineering **323** (2018) 012009 doi:10.1088/1757-899X/323/1/012009

Enhanced oil recovery by nitrogen and carbon dioxide injection followed by low salinity water flooding for tight carbonate reservoir: experimental approach

Essa Georges Lwisa and Ashrakat R Abdulkhalek

United Arab Emirates University, Al Ain, UAE

*Email: Essa.lwisa@uaeu.ac.ae

Abstract. Enhanced Oil Recovery techniques are one of the top priorities of technology development in petroleum industries nowadays due to the increase in demand for oil and gas which cannot be equalized by the primary production or secondary production methods. The main function of EOR process is to displace oil to the production wells by the injection of different fluids to supplement the natural energy present in the reservoir. Moreover, these injecting fluids can also help in the alterations of the properties of the reservoir like lowering the IFTs, wettability alteration, a change in pH value, emulsion formation, clay migration and oil viscosity reduction. The objective of this experiment is to investigate the residual oil recovery by combining the effects of gas injection followed by low salinity water injection for low permeability reservoirs. This is done by a series of flooding tests on selected tight carbonate core samples taken from Zakuum oil field in Abu Dhabi by using firstly low salinity water as the base case and nitrogen & CO_2 injection followed by low salinity water flooding at reservoir conditions of pressure and temperature. The experimental results revealed that a significant improvement of the oil recovery is achieved by the nitrogen injection followed by the low salinity water flooding with a recovery factor of approximately 24% of the residual oil.

1. Introduction

Enhanced Oil Recovery (EOR) refers to the process of producing liquid hydrocarbons by methods other than the conventional use of reservoir energy and reservoir pressurizing schemes with water and gas [1]. This involves the injection of different fluids in order to achieve two main objectives. The first objective in to supplement the natural energy in the reservoir; second to create favorable conditions by the interacting with the rock/oil system for increasing the residual oil recovery by reducing the interfacial tension between the displacing fluid and oil, increase the capillary number, decrease the capillary force, reduce oil viscosity and alter the wettability of reservoir rock [2]. EOR are divided into two major techniques: thermal and non-thermal recovery. Among the non-thermal techniques is the gas flooding and low salinity water flooding. Based on the previous studies low salinity water has been a very effective method in increasing the oil recovery for different reservoir conditions. Moreover, among the non-thermal EOR techniques gas injection particularly carbon dioxide and nitrogen injection are proven to be one of the most attractive and effective methods for increasing the residual oil recovery [2][3].

Macroscopic and microscopic sweep efficiencies are one of the factors that affects the overall efficiency of any EOR process. Macroscopic efficiency is influenced by the difference of densities and rock heterogeneity, whereas the microscopic efficiency is affected by the interfacial interactions which include interfacial tension and dynamic contact angles [1][4]. The residual oil saturations in gas swept zones has been found to be quite low. However, the volumetric sweep of the flood has always been a cause of concern. Since the viscosity of the injected phase is relatively low, the mobility ratio, which controls the volumetric sweep, between injected gas and displaced oil bank in gas processes, is typically highly unfavorable. This difference makes the mobility and consequently the flood profile controls the biggest concern for the successful application of this process [5]. Regarding this concern, the objective of this work is to determine the recovery of oil by the injection of gas followed by the low salinity water flooding, this can result in a significant increase in oil recovery since the higher microscopic displacement efficiency of gas followed by the better macroscopic sweep efficiency of water has been found to significantly increase the incremental oil production over the plain water flood or even plain gas injection. The idea of this experiment is similar to the WAG (water- alternating -Gas) process but in this experiment the gas is separately injected first followed by the injection of low salinity water separately. Moreover, based on the previous studies each injection fluid has an influence on both the oil and rock properties that will be explained as follows:

1.1. Low Salinity Water Flooding

Over the past decade low-salinity water flooding (LSWF) has emerged as a viable enhanced oil recovery method. All the laboratory tests and field trials have proved that the injection of modified salinity water instead of sea water can lead to significant oil recoveries [6]. This is due to the increase of the macroscopic and microscopic sweep efficiency by the lowering and optimization of the salt content of the injected water [7]. Based on the research work that is published there are some conditions that may be necessary to increase the oil recovery by LSWF.

A. Many researches showed that the presence of clay is important during LSWF. Kaolinites and Illites are non-swelling clays that tend to detach from the rock surface and migrate when the colloidal condition are conductive for release. However, different authors reported the importance of different types of clay.
B. Presence of polar oil.
C. Presence of connate water saturation.
D. Salinity shock – the salinity of the injected water must be significantly lower than the preceding water salinity. However, some researches had shown low oil recovery with the presence of all the above condition and other researches had shown high oil recovery with the presence of few conditions, therefore all these conditions cannot guarantee the increase of the efficiency of LSWF [8].

1.2. Carbon dioxide injection

Carbon dioxide has the property if mixing with the oil to swell it, make it lighter, detach it from the rock surfaces, and causing the oil to flow more freely within the reservoir so that it can be "swept up" in the flow from injector well to producer well [2]. The capacity of CO_2 to vaporize hydrocarbons is much greater than that of natural gas. It has been shown that CO_2 vaporizes hydrocarbons primarily in the gasoline and gas-oil range. This capacity of CO_2 to extract hydrocarbons is primary reason for the use of CO_2 as an oil recovery agent. Moreover, previous studies have showed that there is a correlation among the API gravity of the crude oil, the temperature of the reservoir and the minimum miscibility pressure. Further studies revealed that the molecular weight of the C_{5+} fraction was a better variable to use than the API gravity.

1.3. Nitrogen injection

Nitrogen is an inert gas that becomes miscible at special conditions of very high pressure and efficiently reduces the oil viscosity and provides efficient miscible displacement. The past studies proved that nitrogen injection could recover up to 45-90% of initial reserves [2]. Two recovery mechanisms are generally considered with nitrogen injection: miscible and immiscible displacement. In the miscible process, increased recovery is caused by transfer of light components from the oil into the gas phase, where the resulting gas develops miscibility with oil [3]. Moreover, with nitrogen injection there will be vaporizing-gas process. Vaporizing-gas drive miscibility is achieved by vaporization of light hydrocarbon components from the reservoir fluid into the driving gas [3]. Furthermore, nitrogen injection can be an effective oil recovery technique in low permeability reservoirs due to its low molecular mass and small molecular size which makes it able to flow through the very small pores of the reservoir.

2. Experimental fluids

The cores were initially saturated with dead crude oil taken from Bu Hassa field in Abu Dhabi, UAE. This crude oil is classified as light crude oil since its API gravity is 35.7°. Table 1 shows the compositional analysis of the crude oil. Synthetic formation brine was used as the aqueous phase with the composition shown in table 2. The cores are then flooded with Sea water (50,000 ppm of NaCl) to bring the oil saturation to the residual saturation (Sor-1). For the EOR process low salinity water that has concentration of 5,000 ppm of NaCl is used. This low salinity water is prepared using sea water and deionized water from the UAE laboratories. Pure CO_2 and N_2 gases are also used for the EOR process.

Table 1. Compositional analysis of crude oil.

Component	Mole %
H_2	0
H_2S	0
CO_2	2.59
N_2	0.12
C1	34.16
C2	6.72
C3	6.36
iC4	1.54
nC4	4.05
C5	0.01
iC5	1.99
nC5	2.66
C6	4.78
C7	3.82
C8	6.11
C9	2.58
C10+	22.51
Total	100

Table 2. Chemicals concentration in formation water.

Chemicals	gm/ liter
$NaHCO_3$	0.46
Na_2SO_4	1.31
NaCl	111.12
$CaCl_2$	38.32
$MgCl_2\ 6H_2O$	13.42

3. Experimental setup

It consists of a hydrostatic core holder which has two inlet lines: one is connected to a pressure syringe pump that is filled with low salinity water and the other is connected to either nitrogen or carbon dioxide cylinder. A relief valve was connected to the end stem (outlet line) to maintain backpressure at

150 psi. The inlet, back and overburden pressures were measured using pressure regulators mounted on the core flood apparatus. The porous media in this experiment is a selected tight short carbonate core samples.

4. Experimental Procedure

The first step of this experiment is core preparation before starting the EOR process. The core preparation process includes: core cleaning using hot Soxhlet method, measurements of pore volume and permeability by nitrogen, and saturating of the core formation brine. The cores were then brought to connate water saturation (Swi) by oil flooding process. The secondary recovery process in this work is represented by flooding the core samples with Sea water to bring the oil saturation to residual saturation (Sor-1). Finally, the enhanced oil recovery techniques were conducted.

The first technique was by flooding the core with low salinity water as the base case of this experiment and the residual oil produced was measured to calculate the recovery. The second test done was by injecting 5 cc of nitrogen and then flooding the core by 60 cc of the same low salinity water. Finally, the last test was by injecting 5 cc of CO_2 followed by the injection of 60 cc of low salinity water.

5. Results and discussion

The first step of the experiment is flooding the core samples by sea water which to achieve the residual oil saturation. This step represents the secondary oil recovery scheme. Table 3 shows the results of oil flooding phase of the experiment to achieve the connate water saturation. Table 4 shows summary of results for the sea water flooding phase of the experiment which represent the secondary oil recovery for each core sample and the residual oil saturation for each core.

Table 3. Summary of oil flooding tests.

Sample ID	Pore volume (cc)	Porosity (%)	Permeability (mD)	Water produced (cc)	Swi (%)
2	12.9	16.6	0.3	11.6	10.0
4	16.2	19.9	0.5	12.8	21.1
5	19.3	23.5	2.2	12.5	35.3

Table 4. Summary of sea water flooding tests.

Sample ID	Water injected (cc)	Oil produced (cc)	Residual oil (cc)	Sor-1 (%)	Oil recovery (%)
2	60.0	7.4	4.2	36.2	63.8
4	60.0	7.8	5.0	39.1	60.9
5	60.0	8.4	4.1	32.8	67.2

The results of the flooding tests have shown that nitrogen followed by LSW has the highest recovery which is 24% of the residual oil whereas the recovery of LSW water flooding and CO_2 followed by LSW flooding were 7.14% and 7.32% respectively. This result of nitrogen flooding followed by LSW can be attributed by two reasons. Firstly, nitrogen has good injectivity in low permeability reservoirs [2]. This makes nitrogen injecting more effective than CO_2 in this experiment since very tight cores are used. Nitrogen has smaller molecular weight than CO_2 which enables it to reach the very small pores of the cores which cannot be reached by CO_2. Secondly, the vaporization gas drive that is caused by nitrogen flooding. This drive mechanism causes the vaporization of the light components of the oil (C1 to C6) and hence this phenomenon can be more effective with light oil that has high methane concentration

which is similar to the oil that has been used in this experiment. This observation agrees with the laboratory studies that was done by Koch and Hutchinson in 1958. They found that a high methane concentration in the reservoir fluid improved the attainment of vaporizing-gas-drive miscibility with nitrogen, and speculated that the presence of methane in the reservoir fluid is helpful in volatilizing the

IOP Conf. Series: Materials Science and Engineering **323** (2018) 012009 doi:10.1088/1757-899X/323/1/012009

C2 through C6 fraction of oil [3]. Moreover, the recovery of the CO_2 followed by LSW and LSW flooding tests were almost the same which means that there was no interaction between the CO_2 injected and the oil. Table 5 shows results of the EOR flooding tests.

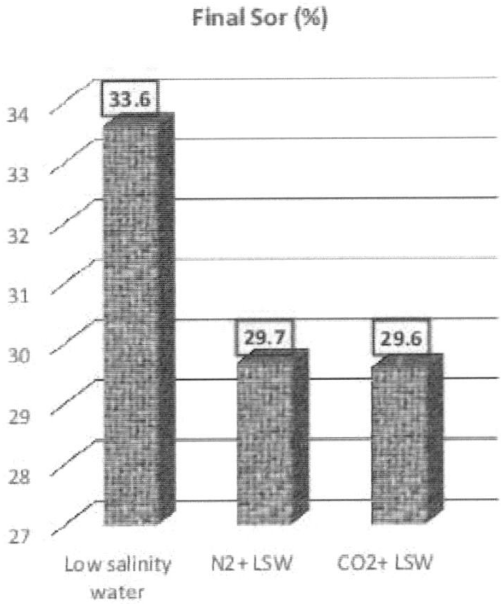

Figure 1. Final oil saturation (%).

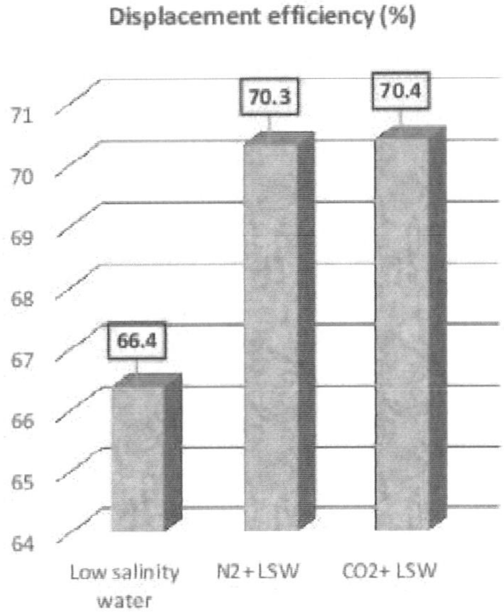

Figure 2. Displacement efficiency (%).

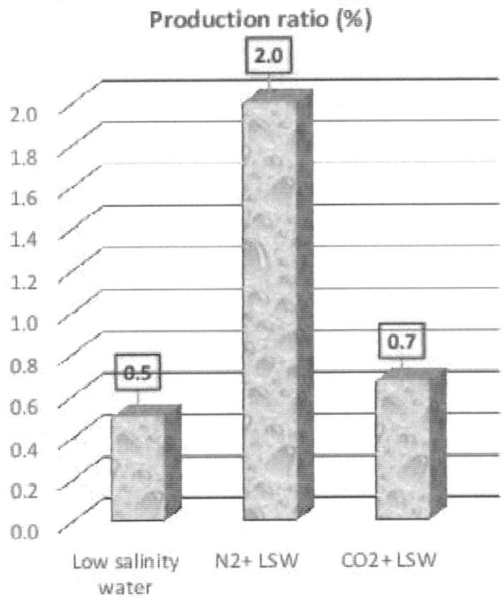

Figure 3. Productio ratio (%).

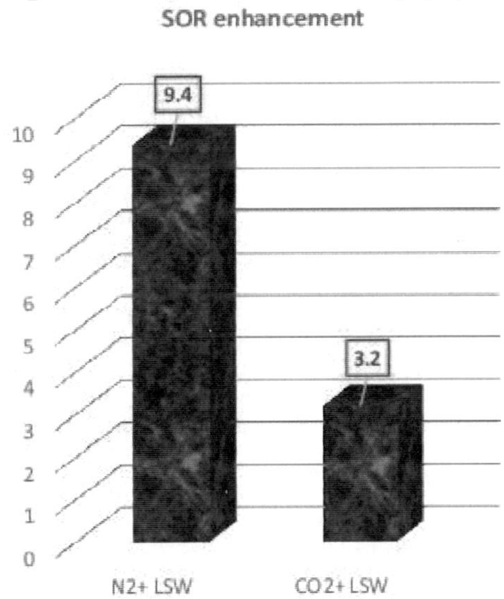

Figure 4. Residual oil saturation. Enhancement.

Table 5. Summary of EOR flooding tests.

Sample ID	Test type	Oil produced (cc)	Sor-2 (%)	EOR (%)	Recovery of residual oil (%)
2	LSW	0.3	33.6	2.6	7.1
4	N2+ LSW	1.2	29.7	9.4	24.0
5	CO2+ LSW	0.4	29.6	3.2	7.3

Figure 1 show the final Residual Oil Saturation which is percentage of volume of residual oil divided by initial volume of oil in core samples. Figure 2 shows the Displacement Efficiency which is percentage of produced oil divided by initial volume of oil. Figure 3 shows the Production Ratio which is the volume of produced oil divided by volume of brine flooded into the sample. Figure 4 shows the enhancement in residual oil saturation.

6. Conclusion

Based on the experimental result of this work, the following conclusions may be drawn:

1. In the conditions of low permeability reservoir, nitrogen and CO_2 injection only result in 0% recovery of residual oil and hence it is more effective to be followed by a liquid fluid in order to benefit from both the interaction of nitrogen and CO_2 with oil and the sweep efficiency of the liquid fluid.
2. In the condition of tight carbonate reservoir and light oil, nitrogen injection followed by low salinity water flooding result in high recovery of residual oil compared with low salinity water flooding and CO_2 followed by low salinity water flooding.

References

[1] Ronald E. Terry," *Enhanced Oil Recovery",* pp. 503-18, 2000.

[2] Abubaker H Alagorni, Zulkefli Bin Yaacob and Abdulrehman H. Nour, "*An Overview of Oil Production Stages: Enhanced Oil Recovery Techniques and Nitrogen Injection",* International Journal of Environmental Science and Development, Vol. **6**, No. 9, 2015.

[3] Oistein Glaso, "*Miscible Displacement: Recovery Tests with Nitrogen",* SPE 17278, 1990.

[4] Willhite, G. P. & Green D. W., "*Enhanced Oil Recovery",* Textbook series, SPE Richardson, Texas, 1998.

[5] Mohamed E Amin, Abdulrazag Y Zekri, Reyadh Almehaidab and Hazim Al-Attar, "*Optimization of CO2 WAG Processes in a Selected Carbonate Reservoir: An Experimental Approach",* Int. Engg. Res. & Sci. & Tech. 2013.

[6] Hazim H. Alattar, Mohammad Y Mahmoud, Abdulrazag Y Zekri, Reyadh Almehaidab, Mamdouh Ghannam, "*Low-salinity flooding in a selected carbonate reservoir: experimental approach",* J Petrol Explor Prod Technol, 2013.

[7] Tibi G. Sorop *et al.*, "*Relative Permeability Measurements to Qualify the Low Salinity Flooding Effect at Field Scale",* SPE 177856-MS, 2015.

[8] J.J. Sheng, "Critical *Review of low salinity water flooding",* Journal of Petroleum Science and Engineering 120, pp. 216-24, 2014.

IOP Conf. Series: Materials Science and Engineering **323** (2018) 012010 doi:10.1088/1757-899X/323/1/012010

Ultrasonically assisted extraction of calcium and ash from char

E.E. Mathumba*, R.K.K. Mbaya and A. Kolesnikov

Department of Chemical, Material and Metallurgical Engineering. Tshwane University of Technology, Private Bag X680, Pretoria, 0001, South Africa

*Email: elvis.mathumba@gmail.com

Abstract. This study characterized and removed calcium and ash content from char to improve the chemical quality of char as reductant for titanium smelting application. Calcium in char can be classified in two parts: mineral matter and cationic metals associated with organic matrix. Virgin and chemically treated char was characterized by using ISO 1171, wet chemistry methods, ISO 19579, XRF, and B.E.T. methods. In this present work, demineralization of char with mild chemical leachants such as acetic acid, citric acid, gluconic acid and Ethylene Diamine Tetra Acetic acid with three different ultrasonic power input (150 W, 270 W and 300 W) and semi-dual frequency of 40 kHz tank was investigated. Actual power dissipated into the system was calculated from the calorimetric measurement. An optimum set of process parameters are identified and validated. The ultrasound technology was compared with soaking technology to determine the efficiency of ultrasound system for the removal of calcium. The removal of calcium was exponentially higher with ultrasonic treatment than without it. Results revealed that mild chemical reagents do not harm the carbon content of char. It is evident from the results that amongst the leachants used; acetic and citric acid has caused significant removal of mineral phases.

1. Introduction

1.1. Background

Char is a solid carbonaceous material produced by heating of low-phosphorus and preferably low-sulphur coal [1]. Coal is heated in a retort at controlled temperature of approximately 900°C, in the absence of oxygen, and char is produced as a combustible residue consisting of fixed carbon and residual ash. In the process, the impurities present in coal, such as volatile matters (tars and oil), are driven off. The process of heating at controlled temperature does not allow burning of carbon present in the coal. High quality of char is commonly utilized as a reductant during the smelting of chromite ores, as well as in the titanium and platinum industries.

1.2. Problem statement

High calcium in char induces limitations on its use for applications of titanium smelting. Titanium is produced by first converting the oxide into the chloride and then react titanium (IV) chloride, ($TiCl_4$) with either sodium or magnesium in the presence of the reductant (char). The chlorination process is sensitive to impurities such as magnesium chloride ($MgCl_2$) and calcium chloride ($CaCl_2$) which affect the stability of the fluidized bed. Large quantities of calcium is deemed as obstacles to the practical use of char because calcium is notorious for causing out of specification product, reducing titanium product quality and therefore the concentrations of magnesium and calcium in reductants have to be low [2]. Therefore, removal of calcium from char prior to its utilization is necessary. Over the years several technologies have been studied for demineralization of carbonaceous material and have been made available [3]-[4]. However, all these methods are tedious, obnoxious and costly. Introduction of less time consuming and cost effective techniques is the need of the day.

1.3. Modern techniques for char-leaching

Very few researchers if not none have focused on ultrasonic char demineralization. Ultrasound is a sound wave with the frequency above the human audible range of 16 kHz. Ultrasound having a frequency range of 20 to 430 kHz is commonly employed for improving the processes such as cleaning, degassing, extraction, etc., for accelerating chemical reactions [5].

The main effects of ultrasonic vibration in liquids are acoustic cavitation and acoustic streaming. Cavitation is due to implosion of bubbles in the acoustic field, and resulting transmission of a shock wave. Whenever ambient pressure is reduced, the boiling point of a liquid is also lowered. If the pressure is reduced far enough, the liquid will begin to boil without needing to be heated (as the boiling point is reduced to below room temperature). When this happens on a small scale, due to localized pressure reduction, small bubbles of vapour are formed [5]-[6]. The subsequent collapse of these bubbles streaming is caused by unidirectional flow currents in a fluid caused by the presence of sound waves. The basic mechanism in effect when ultrasound is coupled to the char slurry is leaching, which removes some of the impurities by a mass transfer. In conventional char demineralisation, the main focus is on surface cleaning, with little penetration to the interior part of the char matrix. Ultrasonic char demineralization is a prime technique for removal of mineral matter from surface and interior part of the char matrix.

In this work, the state-of-the-art ultrasound in char leaching has been studied.

2. Experimental

2.1. Experimental set-up

The ultrasonic system has three components. (i) Ultrasonic generator, (ii) ultrasonic transducer and (iii) tank to contain liquid.

Figure 1. (a) Schematic representation of the experimental set – up.

The ultrasonic tank as shown in figure 1 is a bright-annealed stainless-steel tank that has transducers made of piezo-electric material. The ultrasonic generator transforms the line voltage to a frequency corresponding to the operative frequency of the transducer. The ultrasonic transducer operates at constant frequency of 40 kHz. The Ultrasonic bath has adjustable temperature up to 80°C.

2.2. Materials

The coal and char samples utilised in this study were sourced from a commercial char making operation in South Africa. Some coal and char samples were screened to (-3 +1 mm) and others were unscreened such as char samples produced from char plant (+15 mm and -30 mm). The reagents used were of analytical grade.

2.3. Coal and char properties
Proximate analysis, total sulfur and forms of sulfur of char are given in table 1.

Table 1. Coal and char properties.

	Char	Units
Calcium	0.32 – 1.48	%
Volatile matter (air dried)	2.2 - 6.0	%
Fixed carbon (air dried)	70.7 - 77.7	%
Pyritical Sulphur (air dried)	0.01 – 0.03	%
Sulphatic Sulphur(air dried)	0.03 – 0.09	%
Organic Sulphur(air dried)	0.36 – 0.46	%
Total Sulphur (air dried)	0.40 – 0.58	%
Ash content (dry basis)	18.5 – 22.1	%

2.4. Experimental procedure
As-received coal and char samples were air-dried, some samples were screened and others were unscreened, taken as produced from char plant. Demineralization experiments were carried out by conventional soaking and ultrasonic methods. Three different ultrasonic power inputs (150 W, 270 W and 300 W) and semi-dual frequency of 40 kHz tank were used for this experimental investigation. Acetic acid, citric acid, distilled water, gluconic acid and EDTA acid were used as reagents. The reagents used were of analytical grade. Seventy (70) g of char samples of known size range was placed in a 500ml beaker, and reagent was poured into the beaker, as per the trial condition. The mixture was subjected to sonication. The treated char samples were then washed for about 3 minutes with 70 ml of distilled water and dried for about 10 minutes in an oven at 105°C. The treated char samples were analysed for proximate analysis, forms of sulphur, total sulphur and calcium content, ash composition according to ISO 1171 procedure, wet chemistry methods, ISO 19579 procedure, and XRF procedure.

3. Results and discussion

3.1. Effect of ultrasonic power dissipated on calcium removal
Early investigations by many researches revealed that the mechanism of the collapse of a cluster of cavities is based on the theory of energy transfer [7]-[8] and has continuously been confirmed by later investigations [9], [10]. In the latest review, Huaneng and co-workers [11] summarized the effect of ultrasonic power that when ultrasound is applied to the system, the ultrasonic power accelerates diffusion and increase diffusion coefficient. Ultrasonic power has a significant effect on accelerating the diffusion process, which increases mass deposition rate and make more clusters cross the solid of carbonaceous material and then allow the leachant, to enter the interior of char particles by diffusion through pores. [12]. Calcium removal increased with increasing power dissipated to the bath (see figure 2). Further evidence for this phenomenon is illustrated by the positive correlation (R^2 = 0.9841, with citric acid) observed between calcium removal and dissipated power (figure 2). According to the mass transfer differential equation [13], the variation of liquid velocity could change the distribution of solution concentration. The

study of the change of distribution of solution concentration with the liquid velocity under ultrasound has been reported by Jiao *et al.*, [12]. They reported that the mathematical relationship between mass transfer coefficient and the distribution of solution concentration is affected by liquid velocity. Their study established that mass transfer coefficient increases with the increasing of ultrasound power. Furthermore, in this present work, it was found that calcium removal increases with increasing power dissipated to the bath as shown in figure 2. It follows that 97.32% maximum calcium was removed at 216 W ultrasonic power dissipated using citric acid and 96% of calcium removal was achieved with acetic acid leachant. In 186 W of power dissipated, three different reagents were used for the investigation, *i.e.*, gluconic acid, acetic acid and citric acid. Highest calcium removal of 97.17% was observed in citric acid leachant and interestingly calcium removal of 89.8% was achieved with gluconic acid. In 15.4 W of power dissipated, gluconic acid removed 70.2% of calcium and maximum removal of 84.97% was achieved with citric acid leachant. Higher removal in all cases was observed with citric acid leachant which has high viscosity and density as shown in table 3. This is consistent with observation by Ambedkar [5] which established that once cavitation is initiated, the cavitation effect is stronger in a high-viscosity fluid and consequently increases demineralization.

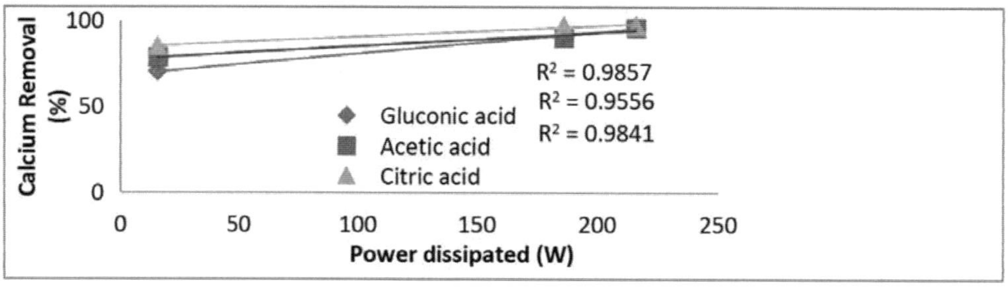

Figure 2. Linear relationship between calcium removal and dissipated power.

3.2. Comparison of sonication and soaking leaching for calcium removal

Figure 3 shows that demineralization of char in reagent ultrasonic leaching is better when compared with soaking char demineralization. The soaking of high-calcium char with EDTA leads to about 56% calcium removal in two days while ultrasonic leaching with the same reagent showed 91.95% calcium removal in 10 minutes. In conventional soaking, the main focus is on surface cleaning, with little penetration to the interior part of the char matrix. The lower extraction caused by conventional soaking might be due to diffusion limitations. The pores of char particles are so narrow or mainly blind that the extraction is unable to admit into these pores, thus causing reduction in its extractability. In reagent ultrasonic char demineralization, the radicals produced by the ultrasound penetrate the pores of the char particles due to micro-streaming effect, and thus removed calcium from surface and interior part of the char matrix.

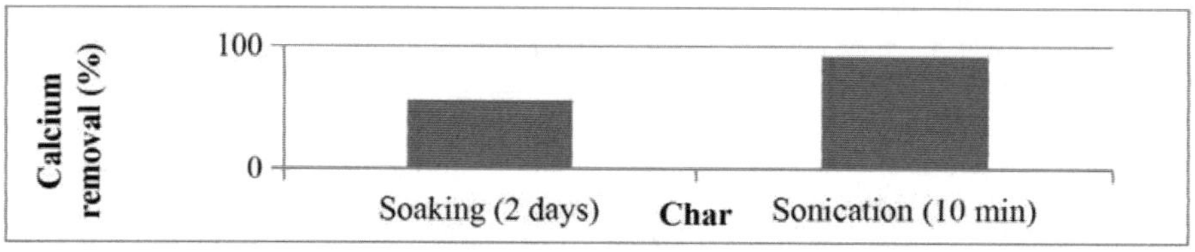

Figure 3. Comparison of sonication and soaking leaching for calcium removal with EDTA.

3.3. Effect on Surface Area

The surface area of char is very important since, like other physico-chemical characteristics, it may strongly affect the reactivity and behavior of char reductant during the smelting of chromite ores, as well as titanium and platinum [10]. Char samples were analysed on a Micromeritics TriStar II surface area and porosity analyser, using Brunauer–Emmett–Teller (B.E.T) method to determine surface area before and after ultrasonic treatment. The B.E.T surface area of char before treatment was 0.35 m^2/g. The results presented in table 2 show that surface area of char material increased when reaction occurred within the system, clearly indicating the more porous char particles than virgin or untreated material. Two leachants were used in this investigation, *i.e.,* citric acid and distilled water. Citric acid created surface area of 37.11 m^2/g, indicating some micro-pore generation. Another factor apparent from the table show that the char particles exposed to sonication had a higher surface area than those exposed to soaking. These results corresponded with the earlier research by Hagenson and Doraiswamy [8].

Table 2. BET Surface area of char before and after treatment.

Sample Name	BET Surface Area (m^2/g)
Virgin Char	0.35
Soaking of char with distilled water (2 days)	1.21
Sonication of char with distilled water (20 min)	1.68
Sonication of char with acetic acid (20 min)	37.11

Table 3. Physical properties of reagents.

	Gluconic acid	Citric acid	Acetic acid
Density (kg/m^3)	1234	1670	1049
Viscosity (kg/m.s) x 10^3	3.02	6.5	1.22

4. Conclusions and recommendations

- Mild leachants such as citric acid and acetic acid may be employed for effective calcium and ash removal from char. Moreover, they do not decrease the carbon content, hence, may be preferred over mineral acids for environmental friendly demineralization of char.
- The soaking of high-calcium char of 1.48% with EDTA leads to about 56% calcium removal in two days while ultrasonic leaching with the same reagent showed 91.95% calcium removal in 10 minutes.
- In almost all cases, higher ultrasonic power dissipated yielded higher calcium removal up to 97.32%. Calcium removal from char was a linear function of sonication time, and thus increased with increasing sonication time.
- Ultrasonic power dissipated of 216 W with acetic acid appears to be optimal for ash content removal up to 43.4%. The ash removal occurred by the larger amplitude of ultrasound traveling through the solvent. The higher the electrical power input, the more violently the bubbles collapsed, and thus more impurities were extracted from the surface and interior of carbonaceous material.
- Acetic acid was found to be most effective for calcium removal from coal and char up to 98.2%.

Acknowledgment
The authors are grateful to commercial char making operation in South Africa for providing the samples. We thank CSIR management, South Africa for warmly accommodating us to use their facilities to perform experimental work. We also thank NRF South Africa and Mott MacDonald for financial support.

IOP Conf. Series: Materials Science and Engineering **323** (2018) 012010 doi:10.1088/1757-899X/323/1/012010

References

[1] J. H. Slaghuis. Coal Gasification. Coal processing IIIA, Part A, pp. 18-29, 1993.
[2] P.C. Pistorius. Ilmenite smelting: the basics. Journal of South African Institute of Mining and Metallurgy, Vol. **108**, pp. 35-43, 2008.
[3] M. Shakirullah, I. Ahmad, H. Rehman, M. Ishaq, U. Khan, and H. Ullah. Effective chemical leaching and ash depletion of low rank coal with EDTA and citric acid. Jour. Chem. Sco. Pak. Vol. **28** (1): pp. 58-61, 2006.
[4] B. Manoj. Chemical demineralization of high volatile Indian bituminous coal by carboxylic acid and characterisation of the products by SEM/EDS. Journal of Environmental Research And Development, Vol. **6** (3A), pp. 654-58, 2012.
[5] B. Ambedkar, "Ultrasonic Coal-Wash for De-Ashing and De-Sulfurization," Ph.D. dissertation, Department of Chemical Engineering, Indian Institute of Technology – Madras, Chennai, India 2012.
[6] K.M. Swamy, L.B. Sukla, K.L. Narayana, R.N. Kar and V.V. Panchanadikar. Use of ultrasound in microbial leaching of nickel from laterites. Ultrasonics Sonochemistry, Vol **2** (1), pp. 7, 1995.
[7] K.A. Morch. "On the collapse of cavity cluster in flow cavitation". Proceedings of the First International Conference on cavitation and in homogeneities in underwater acoustics. Springer Series in Electro physics, Vol. **4**, Springer, Berlin, pp. 95, 1980.
[8] L.C. Hagenson, and L.K. Doraiswamy. Comparison of the effects of ultrasound and mechanical agitation on a reacting solid-liquid system. Chem. Eng. Sci., Vol. **53**, pp. 157-66, 1991.
[9] P.M. Kanthale, P.R. Gogate, A.B. Pandit, and A.M. Wilhelm. Cavity cluster approach for quantification of cavitational intensity in sonochemical reactors. Ultrasonics Sonochemistry, Vol. **10**, pp. 181-82, 2003.
[10] MATHUMBA, EE. "Upgrading the Chemical Quality of Char by Ultrasonication." M.T. dissertation, Department of Chemical, Material and Metallurgical Engineering, Tshwane University of Technology, Pretoria, South Africa, 2015.
[11] X.U. Huaneng, Z. Yingxin, and H.E. Chaohong. Ultrasonically Assisted Extraction of Isoflavones from Stem of Pueraria Zobata (Willd.) Ohwi and Its Mathematical Model, Chin. J. Chem. Eng., vol. **15** (6), pp. 861-86, 2007.
[12] Q. Jiao, Bayanheshig, X. Tan, and J. Zhu. Numerical simulation of ultrasonic enhancement on mass transfer in liquid–solid reaction by a new computational model. Ultrasonics Sonochemistry, Vol. **21**, pp. 535–41, 2014.
[13] J.R. Welty, C.E. Wicks, R.E. Wilson, and G.L. Rorrer. Fundamentals of Momentum, Heat and Mass Transfer, fifth ed., John Wiley & Sons, New York, 2000.

IOP Conf. Series: Materials Science and Engineering **323** (2018) 012011 doi:10.1088/1757-899X/323/1/012011

Extraction of astaxanthin from microalgae: process design and economic feasibility study

Nancy Zgheib*, Roxana Saade, Rindala Khallouf and Hosni Takache

Faculty of Engineering, Chemical Engineering Department, Holy Spirit University of Kaslik (USEK), Jounieh, Lebanon

*E-mail: nancyzgheib@usek.edu.lb

Abstract. In this work, the process design and the economic feasibility of natural astaxanthin extraction from *Haematococcus pluvialis* species have been reported. Complete process drawing of the process was first performed, and then the process was designed including five main steps being the harvesting process, the cell disruption, the spray drying, the supercritical CO_2 extraction and the anaerobic digestion. The major components of the facility would include sedimentation tanks, a disk stack centrifuge, a bed miller, a spray dryer, a multistage compressor, an extractor, a pasteurizer and a digester. All units have been sized assuming a 10 kg/h of dried biomass as a feedstock to produce nearly 2592 kg of astaxanthin per year. The investment payback time and the return on investment were all estimated for different market prices of astaxanthin. Based on the results the production process was found to become economically feasible for a market price higher than 1500$/Kg. Also, a payback period of 1 year and an ROI equal to 113% was estimated for an astaxanthin market price equal to 6000$/Kg.

1. Introduction

The potential of photosynthetic microorganisms is presented in different domains such as food, feed, pharmaceutical, cosmetic and fuel industries. Nowadays, microalgae are under the spot since they constitute a renewable resource. Cultivating this bioresource would solve some of the limitations in the future of traditional biomass production and markets. These microorganisms are known for their high growth rate and biomass productivity compared to vegetables [1]. Recently, microalgae have become the focus of the research on photosynthetic microorganisms due to several reasons. Firstly, these microorganisms are tiny cells that possess different types of pigments responsible of the photosynthesis process. Throughout their growth, pigments, proteins, lipids, and carbohydrates are formed and accumulated in their cells. To grow, they need light, carbon dioxide, nutrients and water or waste water. Consequently, they can be used to reduce the carbon dioxide present in the atmosphere and to lower the concentrations of nitrogen and phosphorus in waste water. Further, due to the limitation in fossil fuels that the world could face, microalgae could be a promising solution for the production of biofuel [2]. The main feature of growing microalgae is the natural commercial extracts obtained after processing them. Of the natural extracts produced from the microalgae *Haematococcus pluvialis*, astaxanthin have received the most attention. The most common use of commercially produced astaxanthin is in fish farming where it imparts coloration to farmed salmonids and crustaceans [3]. Also, due to its antioxidant activity and fortification astaxanthin has been widely used in the pharmaceutical and nutraceutical industry [4]. Nowadays, the commercial astaxanthin is dominated by the synthetic one since the market prices for natural astaxanthin are higher compared to the synthetic version [5, 6]. However, growing consumer demand in the health food sector for natural products has given the opportunity for production of astaxanthin from natural sources. The production of natural astaxanthin being an expensive process, the objective of this work is to develop a new bioprocess to produce natural astaxanthin in a low cost location such as Lebanon and using the available technologies. In this study, we will not include the cultivation process of the biomass rather we will focus on the astaxanthin extraction process. We will assume that the biomass suspension rich in astaxanthin is supplied from a platform producing the microalgae *Haematococcus pluvialis* on a large scale. The purpose of this paper is to explore whether natural astaxanthin production in Lebanon is economically feasible and if the application of the process on an industrial scale could be recommended. This study will be divided into three parts: process synthesis, process design and economic analysis.

2. Description of the production process

The production facility will be theoretically located in Bekaa (Lebanon), a place with considerable sunshine time. The biomass suspension flow rate was estimated to be 10 kg/h of dried biomass as a feedstock to produce nearly 2592 kg of astaxanthin per year. The most important mass inflow refers to freshwater and the suspended microalgae concentration in the inlet flow mixture was considered to be 1 kg/m^3. The main processing units would include two sedimentation tanks, a disk stack centrifuge, a bed miller, a spray dryer, a multistage compressor, an extractor, a pasteurizer and a digester. A continuous process flow diagram of the astaxanthin production was developed as shown in figure 1. According to this study, the harvesting process was performed in a two-step approach employing two gravity sedimentation tanks V-101/102 as a first step followed by a disk-stack centrifugation V-103. Large amount of water was eliminated by sedimentation followed by centrifugation where the concentrated slurry was further dewatered to a cake consisting of 21% of the total suspended solid (TTS). Then, the cake was introduced into a bead miller V-104 in order to disrupt the cell and release the compounds. After harvesting and cell disruption a spray dryer V-105/106 was used to remove the moisture left in the biomass and to increase its shelf life. During this process the algal slurry is dried to 5% moisture content. For the recovery of astaxanthin from *Haematococcus pluvialis*, supercritical CO_2 extraction was used. Throughout this process, the biomass or the paste was entered into the extractor column T-101 where it was mixed with the supercritical CO_2 fluid and the ethanol as a co-solvent at a temperature of 60°C and a pressure of 30 MPa. Since the gas holder tank TK-101 would contain CO_2 gas at 45 bars and 20°C a two-stage compressor C-101 and C-102 and an intercooler E-101 were used to increase the pressure of the CO_2 to 30MPa. Then, a cooler E-102 was used to cool the CO_2 supercritical fluid to 60°C before entering the extraction column T-101. The outlet extract stream was passed through a cyclone V-107 where the astaxanthin was separated from the CO_2 and collected in the tank TK-103. The fluid CO_2 that leaves the cyclone was recycled back to the multistage compressor after being compressed C-103 to 45 bars. The residual biomass (Stream 22) left in the extraction chamber was heated in exchanger E-201 to 65° before entering the pasteurizer E-202. In the pasteurizer the pasteurization of the waste solids was retained at 70° C for 30 minutes to reduce the pathogen levels. The stream 27 was then cooled in exchanger E-203 before entering the anaerobic digester R-201 where the waste was transformed into biofertilizer and biogas. Material balances and process parameters of the astaxanthin production from *Haematococcus pluvialis* were summarized in table 1.

IOP Conf. Series: Materials Science and Engineering **323** (2018) 012011 doi:10.1088/1757-899X/323/1/012011

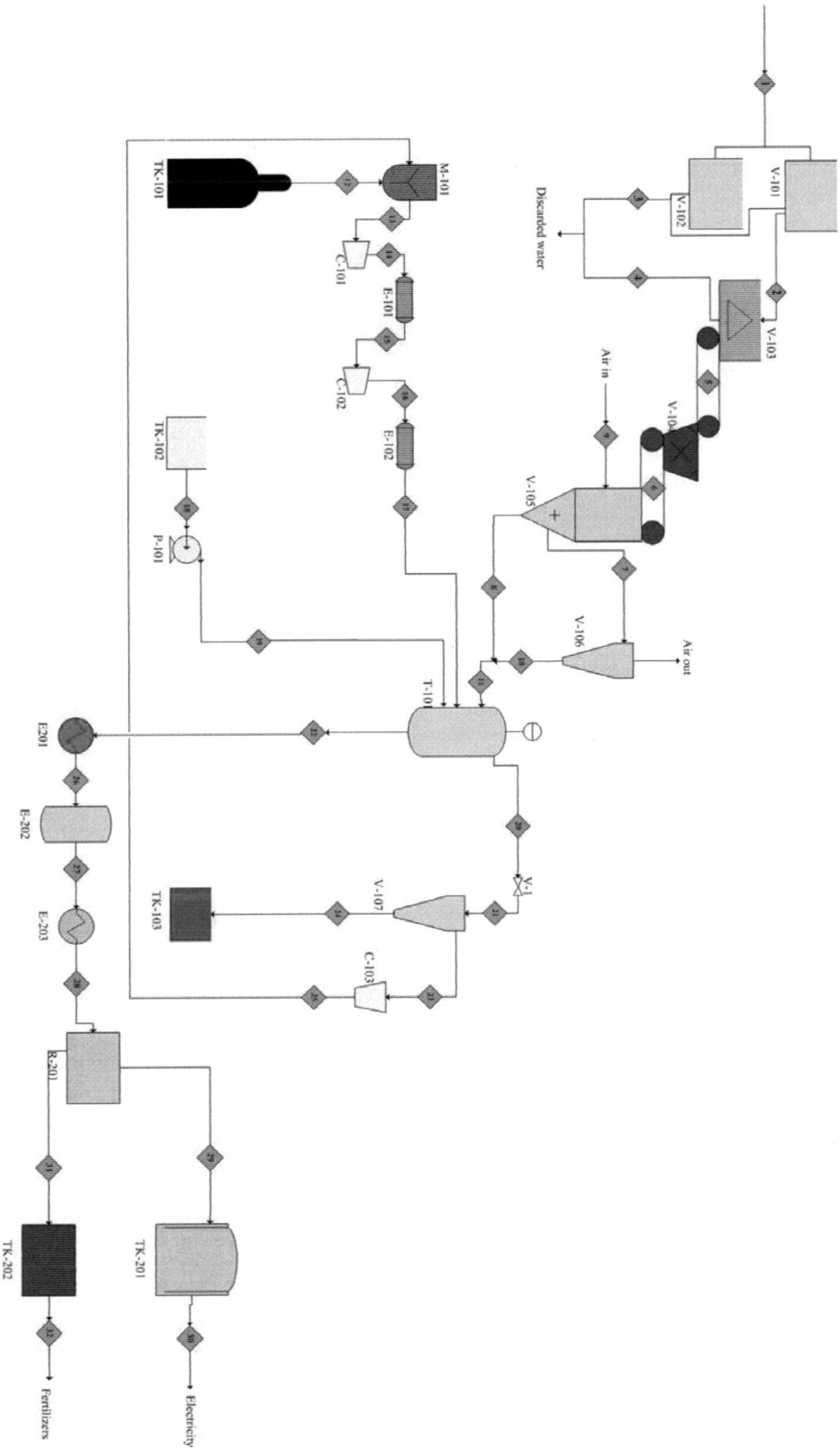

Figure 1. Process flow diagram of the astaxanthin production from *Haematococcus pluvialis*.

Table 1. Material balances of the astaxanthin production from *Haematococcus pluvialis*.

Stream N°	Temperature (°C)	Pressure (bar)	Algae suspension (kg/h)	Air (kg/h)	CO_2 (kg/h)	Ethanol (kg/h)	Astaxanthin (kg/h)	Byproducts (kg/h)	Biogas (m^3/h)	Fertilizer (kg/h)
1	20	1	11,000	-	-	-	-	-	-	-
2	20	1	750	-	-	-	-	-	-	-
3	20	1	10,250	-	-	-	-	-	-	-
4	20	1	703	-	-	-	-	-	-	-
5	20	1	46.8	-	-	-	-	-	-	-
6	20	1	46.8	-	-	-	-	-	-	-
9	35	1	-	12,991	-	-	-	-	-	-
11	20	1	11.8	-	-	-	-	-	-	-
12	20	45	-	-	28	-	-	-	-	-
13	20	45	-	-	561	-	-	-	-	-
14	144	173.5	-	-	561	-	-	-	-	-
15	76	172.5	-	-	561	-	-	-	-	-
16	110	301	-	-	561	-	-	-	-	-
17	60	300	-	-	561	-	-	-	-	-
18	20	1	-	-	-	31.56	-	-	-	-
19	20	1	-	-	-	31.56	-	-	-	-
20	60	300	-	-	-	-	-	-	-	-
21	20	1	-	-	-	-	-	-	-	-
22	40	1	-	-	-	-	-	11.5	-	-
23	20	1	-	-	533	-	-	-	-	-
24	20	1	-	-	-	-	0.3	-	-	-
25	20	45	-	-	533	-	-	-	-	-
26	65	1	-	-	-	-	-	11.5	-	-
27	70	1	-	-	-	-	-	11.5	-	-
28	30	1	-	-	-	-	-	11.5	-	-
29	37	1	-	-	-	-	-	-	6	-
31	37	1	-	-	-	-	-	-	-	8.1

3. Construction of the process

3.1. Harvesting phase

To date, there is no optimal way of harvesting microalgae. The choice of the best process depends on microalgae cell densities, algae species, the growth rates of microalgae, the acceptable level of moisture in the product and the production cost [7–11]. Taking into consideration the average size of a red cell of *Haematococcus pluvialis* (*i.e.* 20 μm) and the concentration of suspended microalgae (*i.e.* 0.1% dry biomass) a two-step approach was used for the harvesting process where the disk-stack centrifugation is combined with the gravity sedimentation technique. Centrifugation is considered an efficient and reliable technique for microalgal recovery especially for concentrating algal slurries from 1% to 5% solids to >15% solids. However, this technique is too cost and energy intensive to be applied alone, therefore, it is recommended as a second stage dewatering technique after a gravity sedimentation step [8, 12, 13, 14]. In this study two sedimentation tanks with a cylindrical shape were considered. Each day the slurry from the algaculture facility would be pumped to these tanks. The sedimentation tanks were sized using Stokes law to allow the maximum possible collection efficiency for a residence time of 1 day. To further increase the biomass recovery (>95%) a disk-stack centrifuge with a relative centrifugal force of 13000 g is used. For mass balance calculations the biomass recovery efficiency was considered to be 95%. The disk-stack centrifuge was sized using the $\sum$ concept, $\sum$ being the equivalent size of the tank in m^2.

3.2. *Cell disruption phase*

Bead milling is a mechanical method already used in industries for cell disruption and have been successfully applied for microalgae [15]. By colliding with each other, the beads with a spherical form and a diameter ranging from 1 to 5 mm will trap and stress the biomass cell which result in breaking the algal cell and releasing the compounds. The bead milling process is a complex operation with many parameters that interact with each other. This interaction is difficult to predict and vary in function of microorganisms, medium viscosity and flow rate. The main operating parameters are the bead filling, bead density, bead diameter, the speed of the agitation and the feed rate [16]. Taking into consideration the high viscosity of the slurry entering the bead miller, ceramic beads specifically alumina (ρ = 3,8 kg/m^3) were used in this study [17]. The optimal bead diameter and the volume fraction of bead loading were obtained from the literature and they were respectively: 0.5 mm and 60% [6]. Additional factors contributing to the milling performance such as the stress number (*SN*) and the stress intensity (*SI*) were calculated using the stress model developed by Kwade and Schwedes [18]. This model which is based on stress energies and bead collision frequencies in the field of microorganisms describes the cell disruption process using two numbers: *SN*, the stress number which is the number of media contact and *SI*, the stress intensity of these events. The *SN* and the *SI* can be calculated as:

$$SN = \frac{\varphi_{\mathrm{GM}}(1-\varepsilon)}{\{1-\varphi_{\mathrm{GM}}(1-\varepsilon)\}C_{\mathrm{V}}} \times \frac{n \times t}{d_{GM}{}^2} \tag{1}$$

$$SI = SI_{\mathrm{GM}} = d_{\mathrm{GM}}{}^3 \times \rho_{\mathrm{GM}} \times v_t{}^2 \tag{2}$$

Where φ_{GM} is the grinding media filling ratio, ε is the bulk porosity, d_{GM} is the bead diameter (m), C_{V} is the cell concentration (kg/m^3), ρ_{GM} is the grinding media density: alumina (kg/m^3), v_t is the grinding media velocity (m/s), n is the agitator speed (revolutions) (m/s) and t is the comminution time: time required to disrupt the cells. For mass balance calculations the biomass recovery efficiency was considered to be 100%.

3.3. *Drying phase*

In order to have high efficiency, the biomass must be dried first, since the moisture can decrease the yield of the supercritical CO_2 extraction by playing the role of a barrier facing the diffusion of the CO_2 into the biomass. The mass flow rate of the air entering the dryer was determined from the material balance performed on the dryer and its value is equal to 12,991 kg air/h. The evaporative rate which is the rate of water evaporated and removed corresponds to 26.125 kg/h and the moisture content in the biomass corresponds to 5%. The residence time in the evaporator was specified to be 2 minutes according to heuristics [19]. The dryer and the cyclone were designed to be a cylindrical chamber with a diameter similar to their height and a 60° conical bottom.

3.4. *Extraction phase*

As soon as the biomass is fully dried the recovery of the astaxanthin is possible. Supercritical CO_2 extraction is widely used to extract high value products from microalgae such as astaxanthin [20, 21]. The fundamental principle behind this method is the use of a supercritical fluid whose properties combine the solvation property of liquids with the high diffusivity of gases [22]. Carbon dioxide is an ideal component for this type of process. Besides being nontoxic, cheap and inert it possesses low critical temperature and pressure (31.1°C and 7.4 MPa). Also, under normal temperature and pressure

CO_2 is a gas which allows the elimination of the solvent after the extraction [20, 23, 24]. In furtherance of enhancing the solvating characteristic of CO_2, the extraction is usually performed using CO_2 as fluid above its critical points at a temperature of 60°C and a pressure of 30 MPa while using ethanol as co-solvent [25]. The rate of CO_2 used in this study is 0.2 kg/min/kg of dry microalgae and the percentage volume of ethanol/CO_2 is 4%. The supercritical CO_2 extraction performed in our study is composed of several units: a CO_2 gas holder tank, a two-stage compressor, a single stage compressor, two heat exchangers, a pump, an extractor column and a cyclone. The two-stage compressor is used to draw the CO_2 off a storage tank and boost the pressure to 30 MPa. Two-stages were used in this study since the compression ratio is equal to 6.66. The compressor utilized is a reciprocating compressor with an efficiency of 80% [26]. The discharge temperature from the first stage being high (T = 145.5 °C), an intercooler was introduced between the first and the second stage. Also, another intercooler was considered between the compressor and the extractor since the temperature of the CO_2 in the extractor should not exceed 60°C. The two-stage compressor was designed by considering that the total work for the system is the sum of the work requirement for the two compressors. To validate our calculations Aspen Hysys was used (See figure 2) assuming an isentropic compression with an efficiency of 80% and employing the SRK (Soave Redlich Kwong) method since the working fluid CO_2 is non-polar [27]. The extraction chamber where the microalgae is placed for subjection to supercritical fluid is cylindrical in shape and closed with stainless steel porous discs. The extraction was carried out at 60°C and 300 bars during 4h. For mass balance calculation we considered that the efficiency of the extraction is 97% and that according to literature [6] 2.5% of astaxanthin are present in the biomass. ProRox SL 930 (alumina and glass) were used for the insulation with a thermal conductivity of 0.038 W/m. The thickness of the insulation was calculated as being equal to 0.0078 m. The cyclone used to separate the astaxanthin from the CO_2 was designed by considering an inlet flow velocity of 0.4 m/s.

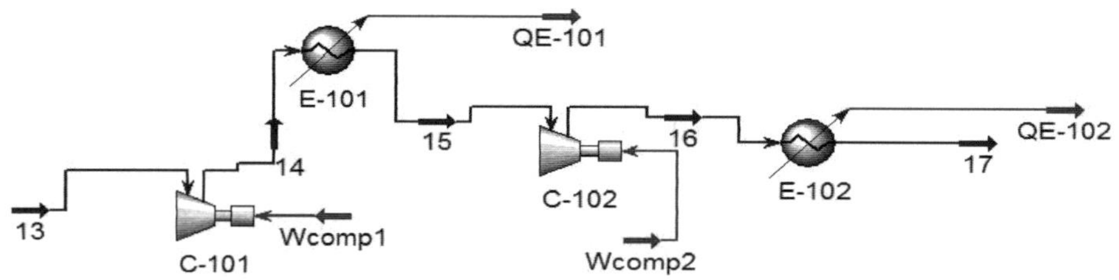

Streams	13	14	15	16	17
Temperature (°C)	20	143.6	75.64	110.1	60
Pressure (bar)	45	173.5	172.5	301	300
Mass flow rate (kg/h)	561	561	561	561	561

Figure 2. Aspen HYSYS Simulation Flow Sheet and stream conditions of the two-stage compressor.

3.5. Residual biomass treatment phase

The residual biomass left in the extraction chamber is rich in carbohydrates and could undergo further processing in order to obtain more valuable products. Anaerobic digestion is one of the most cost effective and environmentally friendly approaches used to treat the residual microalgae. The integration of anaerobic digestion with astaxanthin production improves sustainability in the production process. This process is a natural biological process that stabilizes in the absence of air the waste as organic waste and transforms it into biofertilizer and biogas. To reduce pathogen levels in the residual solids obtained from anaerobic digestion, a pasteurization step was performed prior to the anaerobic process. During pasteurization the waste solids were retained at 70°C for 30 minutes and steam injection was used to heat the viscous influent slurry. The digester was designed for a 21 day retention time and as having a cylindrical shape with a height equal to its diameter. ProRox SL 930 (alumina and glass) were used as insulation with a thermal conductivity of 0.038 W/m.K [28]. The thickness of the insulation was calculated as being equal to 0.003 m. Most of the biogas was produced in the digester at atmospheric pressure and 35 to 37°C [29]. The gas holder volume was assumed to be half of the size of the digester volume. Since the methane is the only component in the biogas with an energy value, the biogas energy value is equal to that of methane. Knowing that 1m^3 of methane contained in the biogas produce about 6 kWh of electricity, the quantity of methane produced was used to power some equipment of the plant. The biomass produced was used as a soil fertilizing amendment.

4. Process design

Using the principles defined in the literature [30– 38], the equipment were designed accordingly. The specifications and performance characteristics of each unit are shown in table 2.

Table 2. Design specification for each equipment of the process.

Code	Name	Capacity/size	Quantity
V-101/V-102	Sedimentation tank	• Volume of 180 m^3	2
V-103	Disk-stack centrifuge	• Length of 0.27 m • Area of 3.24 m^2 • Bowl diameter of 10 inch • Speed of 9500 RPM • Biomass recovery efficiency 95%	1
Stream 5	Conveyor belt	• Width of 0.6 m • Length of 6 m	2
V-104	Bead miller	• Stress number *SN* of 2.9× 10^7 • Stress Intensity *SI* of 1,043 • Biomass recovery efficiency 100%	1
V-105/V-106	Spray Dryer	• Volume of 442 m^3 • Biomass recovery efficiency 95%	1
M-101	Gas mixer	• Retention time of 10 s • Volume of 0.016 m^3	1
C-101	CO_2 compressor	• Efficiency of 80% • Actual Work of 11.9 kW	1
E-101	Intercooler	• Efficiency of 80% • Area of 3.5 m^2 • Work of 21.1 kW	1
C-102	CO_2 compressor	• Efficiency of 82% • Actual work of 4.48 kW	1
C-103	CO_2 Recycled compressor	• Efficiency of 80% • Actual work of 57 kW	1
E-102	Cooler	• Area of 2.55 m^2 • Work of 15.96 kW	1
P-101	Ethanol pump	• Efficiency of 77% • Actual work of 0.44 W	1
V-1	Valve	• Valve flow coefficient of 1.06	1
T-101	Extractor column	• Volume of 5.6 m^3 • Biomass recovery efficiency 97%	1
V-107	Cyclone	• Cyclone dimension of 0.075 m • Biomass recovery efficiency 95%	1
TK-103	Astaxanthin storage tank	• Volume of 2 m^3	1
E-201	Heater	• Area of 16.95 m^2	1
E-202	Pasteurizer	• Area of 10.96 m^2	1
E-203	Cooler	• Area of 21.28 m^2	1
R-201	Digester	• Retention time of 21 days • Volume of 11.6 m^3 • Diameter = Height = 2.4 m	1
TK201	Gas holder volume	• Volume of 5.8 m^3	1
TK-202	Digestate storage tank	• Volume of 1.5 m^3	1

5. Economic Assessment

To evaluate the project profitability, we performed an economic estimation based on the process flow diagram and a rough sizing of the major equipment. The operation days correspond to 360 day in an annual basis and the cost associated with the services and raw materials are listed in table 3. The total capital investment and the total manufacturing cost are demonstrated in table 4 and table 5.

Table 3. Costs of raw materials and services used in the process.

Item	Price
Chemicals	
CO_2	0.4706 \$/kg
Ethanol	784.76 \$/m^3
Utilities	
Biomass	32.16 \$/kg
Cooling water	0.0294 \$/m^3
Steam	0.0049 \$/kg
Electricity	0.13\$/kWh

Table 4. Equipment costs, fixed capital costs and total capital investment.

Type		Cost (\$)
Vessels	V-101	208,019
	V-102	208,019
Centrifuge	V-103	247,308
Bead miller	V-104	64,800
Spray dryer	V-105 V-106	30,000
Conveyor belt	(x2)	32,000
Mixer	M-101	5,000
Compressors	C-101	305,056
	C-102	305,056
	C-103	295,811
Heat exchangers	E-101	10,650
	E-102	8,275
	E-201	29,612
	E-202	29,955
	E-203	30,118
Pump	P-101	40
Extractor	T-101	36,591
Cyclone	V-107	5,000
Storage tank	TK-103	4,871
	TK-201	4,915
	TK-202	3,777
Reactor	R-201	146,393
Valve	V-1	64,152
Total equipment costs (TEC)		2,075,418
Equipment erection = 0.5*TEC		1,037,709
Piping = 0.6*TEC		1,245,251
Instrumentation and Control = 0.3*TEC		622,625
Electrical = 0.2*TEC		415,084
Civil = 0.3*TEC		622,625
Structures and Buildings = 0.2*TEC		415,084
Lagging and Paint = 0.1*TEC		207,542
Offsite = 0.4*TEC		830,167
Design and Engineering = 0.25*TEC		518,855
Contingency = 0.1*TEC		207,542
Fixed capital costs		6,122,483
Fixed capital investment (FCI)		8,197,901
*Working capital investment (WCI) = 0.15*FCI*		1,229,685
Total capital investment = FCI + WCI		9,427,586

IOP Conf. Series: Materials Science and Engineering **323** (2018) 012011 doi:10.1088/1757-899X/323/1/012011

Table 5. Total manufacturing cost.

Item	Cost ($)
Direct manufacturing cost (DMC)	
Raw materials	
Biomass	2,778,624
CO_2	1056
Ethanol	4520
Utilities	
Cooling water	200
Steam	1587
Electricity	82,331
Labor	237,089
Supervision = 0.2 * labor	47,418
Plant overhead = 0.5 * labor	118,545
Interest = 0.08 * FCI	655,832
Insurance = 0.01 * FCI	81,979
Rent = 0.01 * FCI	81,979
Royalties = 0.01* FCI	81,979
Maintenance = 0.075 * FCI	614,843
Miscellaneous = 0.1 * maintenance	61,484
Shipping and packaging = 0.02 * raw materials	55,684
Subtotal (DMC)	4,905,150
*Indirect manufacturing cost (IDMC) = 0.25 * DMC*	1,226,287
Total manufacturing cost (TMC) = DMC + IDMC	6,131,438

5.1. Total capital investment (TCI)
The TCI is the sum of the fixed-capital investment (FCI) and the working capital (WC). The FCI represents the capital needed to supply the necessary manufacturing and plant facilities and it includes the equipment cost, installation, piping, instrumentation and control, electrical, civil, structures and buildings, lagging and paint, offsite, design and engineering and contingency. The WC refers to the capital for sustaining operations before any product is sold and any income is available.

5.2. Total manufacturing cost (TMC)
All costs related to the manufacturing operations or the physical equipment of a plant are comprised in the manufacturing cost. This cost must be known to judge the viability of a project, and to choose the best alternative processing scheme. The operating costs are divided into two groups: the direct manufacturing cost (DMC) and the indirect manufacturing cost (IDMC). The DMC comprise cost of utilities, raw materials, miscellaneous, shipping and packing, labor, plant overhead, supervision, depreciation, interest, insurance, royalties, rent and maintenance. The operating labor cost is calculated by multiplying the total number of worker by the salary or hourly rate of each worker. The monthly individual labor cost is considered to be 500 $/month. The indirect manufacturing cost (IDMC) includes the distribution, sales, research, general overheads and development and it amounts to 25% of the DMC.

5.3. Return on investment (ROI) and breakeven point
The ROI refers to the return on investment and is used to check the feasibility of the astaxanthin production from a business point of view. The annual profit must be calculated in order to determine the ROI. The ROI is calculated using Equation 3 where CAD refers to the cash available to be distributed as shareholder dividends and it is the subtraction between EBITDA and taxes. EBITDA is the net income before interest, taxes, depreciation, and amortization. Interest on capital during construction period and average feedstock and product inventories are not considered in this production.

$$ROI = \frac{CAD}{FCI} * 100 \quad (3)$$

To determine the point at which the production starts to generate profit to the industry, we calculated the payout time using the equation below:

$$Payout\ time = \frac{Original\ depreciable\ fixed\ capital\ investment}{EAT} \tag{4}$$

where EAT represents the sum of the earnings after taxes and the depreciation for a one-year period. Table 6 shows the ROI and the payout time evaluation for an astaxanthin production from microalgae calculated for different market prices of pure astaxanthin (1500, 3000, 4500, 6000, 7000 $/Kg) . The fertilizer was also exploited with an average market price of 40$/Kg. This table shows that an economic viability of the company could be achieved for an astaxanthin market price higher than 1500$/Kg. The values of ROI were found to increase with astaxanthin market prices to attain 113% for an astaxanthin price of 6000$/Kg. Also, for this market price the cash flow will completely offset the depreciable investment after one year of production.

Table 6. Financial statement of the plant.

Price ($)/Kg	Astaxanthin	1500	3000	4500	6000	7000
Kg	Astaxanthin	2,592	2,592	2,592	2,592	2,592
Price ($)/Kg	Ferlilizer	40	40	40	40	40
Kg	Ferlilizer	69,984	69,984	69,984	69,984	69,984
Gross revenue		6,687,360	10,575,360	14,463,360	18351360	20,943,360
VAT (11%)		735,610	1,163,290	1,590,970	2,018,650	2,303,770
Total revenues		5,951,750	9,412,070	12,872,390	16,332,710	18,639,590
TMC		6,131,438	6,131,438	6,131,438	6,131,438	6,131,438
EBIDTA		-179,688	3,280,632	6,740,952	10,201,272	12,508,152
Depreciation (10%)		819,790	819,790	819,790	819,790	819,790
EBIT		-999,478	2,460,842	5,921,162	9,381,482	11,688,362
Interest expense debt						
Interest income on cash						
EBT		-999,478	2,460,842	5,921,162	9,381,482	11,688,362
TAX (10%)			246,084	592,116	938,148	1,168,836
EAT		-999,478	2,214,758	5,329,046	8,443,334	10,519,526
CAD		-179,688	3,034,548	6,148,836	9,263,124	11,339,316
FCI		8,197,901	8,197,901	8,197,901	8,197,901	8,197,901
ROI (%)		-2,19	37.01	75.01	112.99	138.32
Payout time (years)			3	1.5	0.97	0.78

6. Conclusion

This theoretical study investigated the large scale production of astaxanthin from *Haematococcus pluvialis* for a potential company located in Bekaa (Lebanon). The technical performance of the process was evaluated by sizing the major equipment of the process. Taking into consideration an astaxanthin production of 2592 kg per year, a complete process flowsheet was designed. Detailed operating conditions and equipment designs for the process were obtained. Then, a detailed economic assessment of the project was developed by performing a Profit and Loss analysis. The results have proved the economic feasibility of the production for different astaxanthin market prices. The economic estimation in this paper is classified as a ''study estimate'' since some information like the process instrumentation diagram or piping and instrumentation requirements, were not considered. Finally, this study could be used as a reference or part of another study that might include the process design of the cultivation of microalgae.

References

[1] Bajpai RK, Prokop A and Zappi ME 2014 *Algal Biorefineries - Volume 1: Cultivation of Cells and products* (Netherlands: Springer)
[2] Mata TM, Martins AA, and Caetano NS 2010 Microalgae for biodiesel production and other applications: A review *Renew. Sustain. Energy Rev* **14(1)** pp. 217–32
[3] Koller M, Muhr A, and Braunegg G 2014 Microalgae as versatile cellular factories for valued

products *Algal Res.* **6** pp. 52–63

[4] Cardozo KHM et al. 2007 Metabolites from algae with economical impact *Comp. Biochem. Physiol.* Part C Toxicol. Pharmacol. **146(1-2)** pp. 60–78

[5] Pérez-López P et al. 2014 Life cycle assessment of the production of the red antioxidant carotenoid astaxanthin by microalgae: from lab to pilot scale *J. Clean. Prod.* **64** pp. 332–44

[6] Li J, Zhu D, Niu J, Shen S, and Wang G 2011 An economic assessment of astaxanthin production by large scale cultivation of Haematococcus pluvialis *Biotechnol. Adv.* **29(6)** pp. 568–74

[7] Panis G and Carreon JR 2016 Commercial astaxanthin production derived by green alga Haematococcus pluvialis: A microalgae process model and a techno-economic assessment all through production line *Algal Res.* **18** pp. 175–90

[8] Christenson L and Sims R 2011 Production and harvesting of microalgae for wastewater treatment, biofuels, and bioproducts *Biotechnol. Adv.* **29(6)** pp. 686–702

[9] Mata TM, Martins AA, and Caetano NS 2010 Microalgae for biodiesel production and other applications: A review *Renew. Sustain. Energy Rev.* **14(1)** pp. 217–32

[10] Molina Grima E, Belarbi EH, Acién Fernández F, Robles Medina A, and Chisti Y 2003 Recovery of microalgal biomass and metabolites: process options and economics *Biotechnol. Adv.* **20(7–8)** pp. 491–515

[11] Rawat I, Ranjith Kumar R, Mutanda T, and Bux F 2011 Dual role of microalgae: Phycoremediation of domestic wastewater and biomass production for sustainable biofuels production *Appl. Energy.* **88(10)** pp. 3411–24

[12] Barros AI, Gonçalves AL, Simões M, and Pires JCM 2015 Harvesting techniques applied to microalgae: A review *Renew. Sustain. Energy Rev.* **41** pp. 1489–1500

[13] Show KY, Lee DJ, and Chang JS 2013 Algal biomass dehydration *Bioresour. Technol.* **135** pp. 720–29

[14] Rawat I, Ranjith Kumar R, Mutanda T, and Bux T 2011 Dual role of microalgae: Phycoremediation of domestic wastewater and biomass production for sustainable biofuels production *Appl. Energy.* **88(10)** pp. 3411–24

[15] Günerken E, D'Hondt E, Eppink MHM, Garcia-Gonzalez L, Elst K, and Wijffels RH 2015 Cell disruption for microalgae biorefineries *Biotechnol. Adv.* **33(2)** pp. 243–60

[16] Montalescot V et al. 2015 Optimization of bead milling parameters for the cell disruption of microalgae: Process modeling and application to Porphyridium cruentum and Nannochloropsis oculata *Bioresour. Technol.* **196** pp. 339–46

[17] Shah MMR, Liang Y, Cheng JJ, and Daroch M 2016 Astaxanthin-Producing Green Microalga Haematococcus pluvialis: From Single Cell to High Value Commercial Products *Front. Plant Sci.* **7**

[18] Kwade A and Schwedes J 2007 *Handbook of Powder Technology* (Elsevier) Chapter 6 Wet Grinding in Stirred Media Mills pp. 251–382

[19] Hall S 2012 *Rules of Thumb for Chemical Engineers (Fifth Edition)* (Oxford: Butterworth-Heinemann) Resources pp. 422–28

[20] Mercer P and Armenta RE 2011 Developments in oil extraction from microalgae *Eur. J. Lipid Sci. Technol.* **113(5)** pp. 539–47

[21] Sudhakar K, Premalatha M, and Rajesh M 2014 Large-scale open pond algae biomass yield analysis in India: a case study *Int. J. Sustain. Energy.* **33(2)** pp. 304–15

[22] Mohan SV, Devi MP, Subhash GV, and Chandra R 2014 *Algae Oils as Fuels. Biofuels from Algae* (Elsevier) pp. 155–87

[23] Cooney M, Young G, and Nagle N 2009 Extraction of Bio-oils from Microalgae *Sep. Purif. Rev.* **38(4)** pp. 291–325

[24] Sahena F et al 2009 Application of supercritical CO_2 in lipid extraction – A review *J. Food Eng.* **95(2)** pp. 240–53

[25] Panis G and Carreon JR 2016 Commercial astaxanthin production derived by green alga

IOP Conf. Series: Materials Science and Engineering **323** (2018) 012011 doi:10.1088/1757-899X/323/1/012011

Haematococcus pluvialis: A microalgae process model and a techno-economic assessment all through production line *Algal Res.* **18** pp. 175–90
[26] Hall S 2012 *Rules of Thumb for Chemical Engineers (Fifth Edition)* (Oxford: Butterworth-Heinemann) Chapter 6, Fans, Blowers, and Compressors pp.118–33
[27] Chaves IDG, López JRG, Zapata JLG, Robayo AL, and Niño GR 2015 *Process Analysis and Simulation in Chemical Engineering* (Springer)
[28] Zhang Q, Hu J, and Lee DJ 2016 Biogas from anaerobic digestion processes: Research updates *Renew. Energy.* **98** pp. 108–19
[29] [Online] Available from: http://guide.rockwool-rti.com/products/industry/prorox-sl-930-(rockwool-multiboard).aspx
[30] Hall S 2012 *Rules of Thumb for Chemical Engineers* (Elsevier) Chapter 7, Driver pp. 134–47
[31] Hall S 2012 *Rules of Thumb for Chemical Engineers* (Elsevier) Chapter 6, Fans, Blowers, and Compressors pp. 118–33
[32] Hall S 2012 *Rules of Thumb for Chemical Engineers* (Elsevier) Chapter 2, Heat Exchangers pp. 27–57
[33] Hall S 2012 *Rules of Thumb for Chemical Engineers* (Elsevier) Chapter 5, Pumps pp. 101–117.
[34] Towler G and Sinnott R 2013 *Chemical Engineering Design* (Elsevier) Chapter 7, Capital Cost Estimating pp. 307–54
[35] Towler G and Sinnott R 2013 *Chemical Engineering Design* (Elsevier) Chapter 15, Design of Reactors and Mixers pp. 631–751
[36] Towler G and Sinnott R 2013 *Chemical Engineering Design* (Elsevier) Chapter 19, Heat-Transfer Equipment pp. 1047–1205
[37] Towler G and Sinnott R 2013 *Chemical Engineering Design* (Elsevier) Chapter 18, Specification and Design of Solids-Handling Equipment pp. 937–1046
[38] Perry RH, Green DW and Maloney JO 2007 *Perry's Chemical Engineers Handbook. Eighth Edition* (New York: McGraw-Hill)

IOP Conf. Series: Materials Science and Engineering **323** (2018) 012012 doi:10.1088/1757-899X/323/1/012012

Preparation of Expanded Graphite Using a Thermal Method

Nyssanbayeva Gulnura[1,2,*], Kudaibergenov Kenes[1,2], Ongarbayev Yerdos[1], Mansurov Zulkhair[2] and Roberto Di Capua[3]

[1]Al-Farabi Kazakh National University, Almaty, Kazakhstan

[2]Institute of Combustion Problems, Almaty, Kazakhstan

[3]University of Naples "Federico II", Italy

*E-mail: gulnur.83.29@mail.ru

Abstract. This article reports a new way to prepare expanded graphite using a thermal method. Natural graphite was added to crystalline hydrate of metals with mechanical stirring at room temperature, taken in the number of 20-80% of the mass of mix. This was then placed in muffle furnace. All process of activation takes from 10 to 20 minutes. X-ray diffraction patterns were used to analyze the structure and confirm that expanded graphite had indeed been prepared. A scanning electron microscope was utilized to observe the morphologies of the expandable graphite and expanded graphite.

1. Introduction

The liquidation of oil spills on water surfaces aims to reduce the damage to the ecological and socio-economical resources, while reducing the time required for recovery of these resources and providing acceptable standards of cleanup [1]. Different methods can be used for the removal of petroleum from the water surface, for example, thermal, biological, mechanical and physicochemical (using coagulants and adsorbent materials) techniques [2].

Expandable graphite (EG) is a type of graphite intercalation compound (GIC). Its derivatives are functional carbon materials that can be applied in various fields, such as airtight materials, oil absorbents, flame retardants, high-power batteries, electrodes, and military materials [3-6].

V. Sridhar [7] reported that GICs can be obtained only by one-step at room temperature with intercalation by concentrated sulfuric acid and ammonium persulfate. The resultant EG (denoted as RTEG) exhibits an expanded volume (EV) of up to 225 ml/g. Finally, the excessive usage of concentrated H_2SO_4 as the intercalant in the conventional method can also cause serious environmental pollution [8]. Therefore, it is of great value to seek a simpler and less-polluted approach to prepare EG.

2. Experimental part

In this work are used native graphite and crystalline hydrates of zinc. The graphite from the Zavalye Graphite Plant (Ukraine) is a large-scaly natural graphite subjected to chemical desalination under industrial conditions. The result is achieved by mechanical mixing of powder of initial graphite by the making foam agent by crystalline hydrate of zinc for training of porous structure, taken in the number of 20-80% of the mass of mix. First, NG (2 g) was added to crystalline hydrate of zinc (8 g) with mechanical stirring for 1 min at room temperature. This was then placed in muffle furnace. The muffle furnace was sealed and quickly heated to the desired temperature. Where it was maintained for 20 min.

It was allowed to cool to room temperature naturally. The expanded volume was determined by expanding of the EG and then measuring its volume with a graduated cylinder. So, the experiment proceeds in two stages: 1) mixing of graphite with crystalline hydrate of zinc; 2) and heating of components at a temperature of 350-1000°C. The mix heats up 5-10 minutes. All process of activation takes from 10 to 20 minutes. Scheme of obtaining synthesis of thermally expanded graphite are given in figure 1.

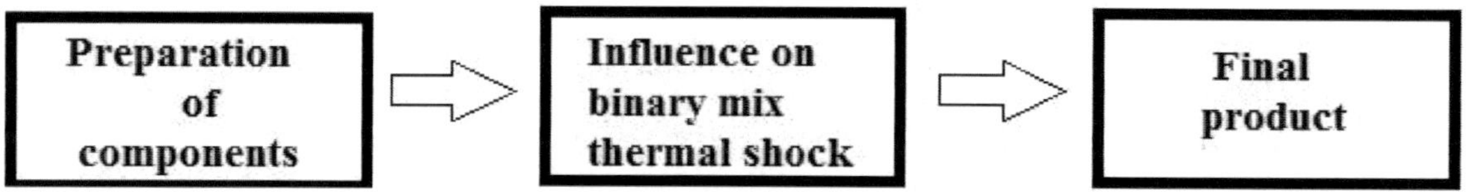

Figure 1. Scheme of the technology of synthesis of the interconnected graphite.

Nature graphite **Expansion process** **Expanded graphite**

Figure 2. Schematically illustration of formation of EG and the synthesis.

In figure 2. shows, how the production of expanded graphite proceeds. On the picture 1 take mix of natural graphite and metal salts. And next picture show expansion process. It is proceeds in a muffle furnace. After 20 minutes we get the finished product. Finished product is expanded graphite.

3. Results and their discussion
The single most striking observation to emerge from the data comparison was formation of the homogeneous melted bubbling mass. After end of this stage there is a sharp foaming of graphite which is followed by allocation of insignificant amount of brown gas (figure 3).

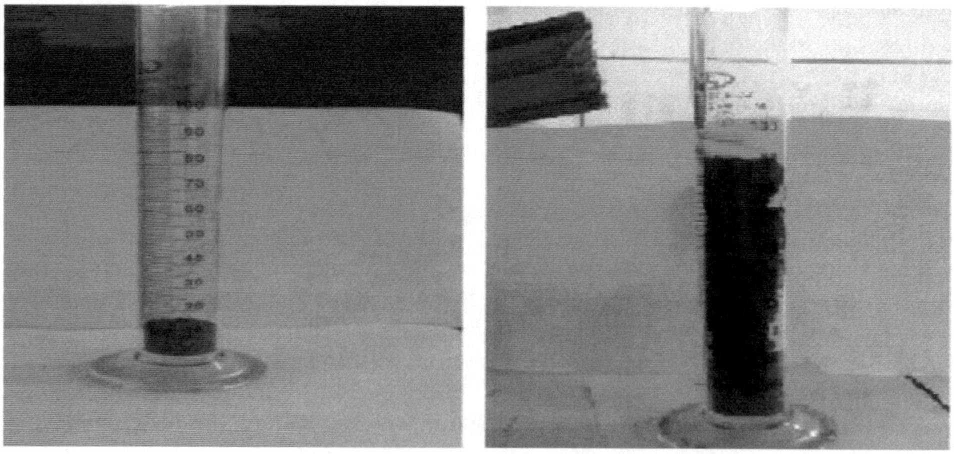

Figure 3. a) mechanical mix before heat treatment b) obtained porous mix after heat treatment.

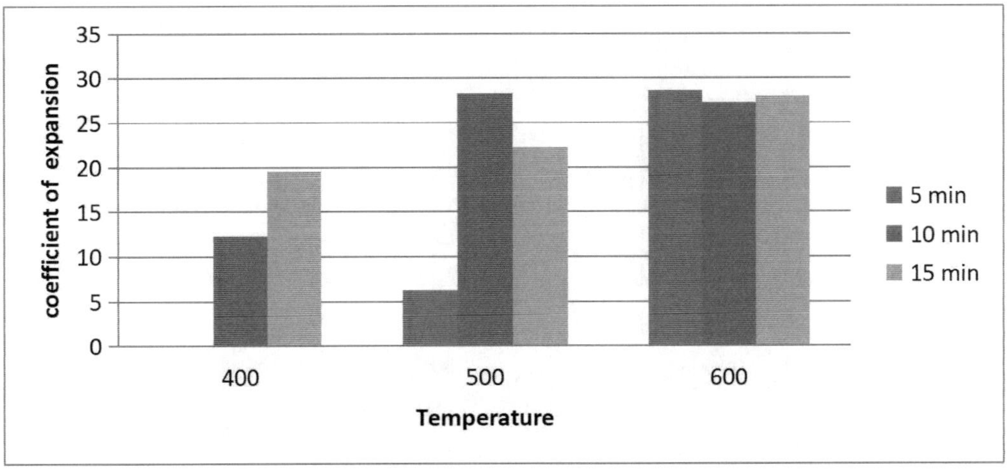

Figure 4. - diagram between coefficient of expansion (C_V), temperature (τ) and time interval.

Figure 4 shows the diagram between coefficient of expansion (CV), temperature (τ) and time interval. It can be explained with the fact that increase in temperature causes thermal decomposition of crystalline hydrate in reagent structure, therefore, the relative content of oxygen increases. Qualitative analyses of reagent structures are submitted in the figure 4.

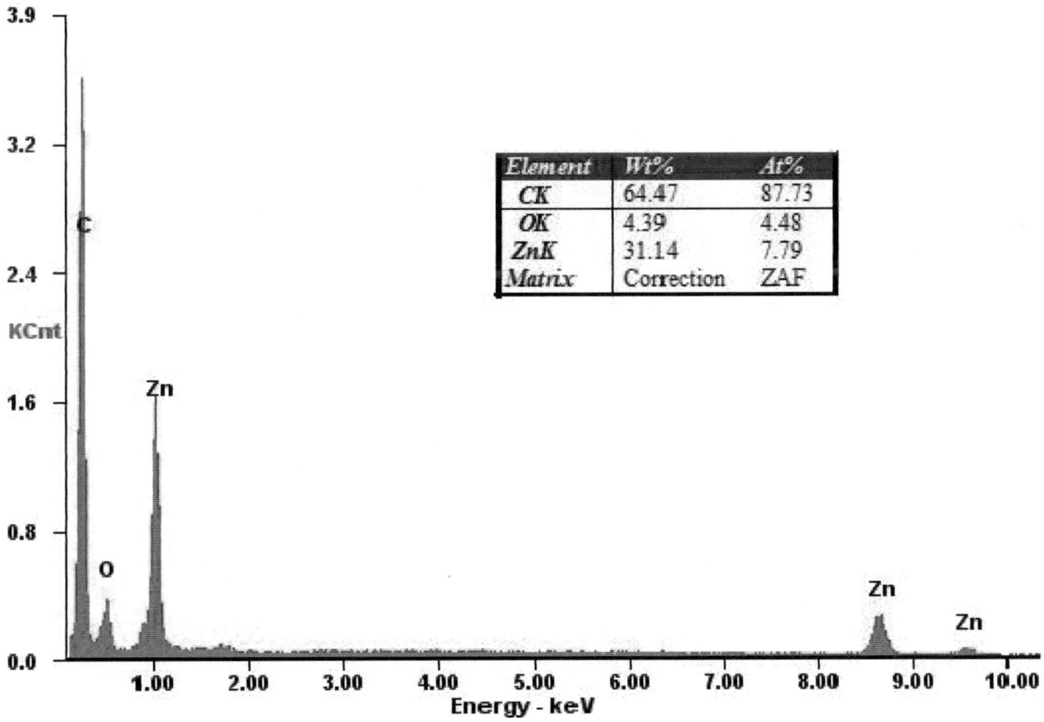

Figure 5. Microanalysis of expanded graphite.

Figures 5 shows the results of microanalyses of expanded graphite using SEM/EDAX. In figures 5 the variation of weight percent of three elements (C, O, Zn) in expanded graphite. Figure 5 also shows that the principal elements in expanded graphite is carbon.

Results microanalysis the thermally expanded graphite obtained from mechanical mix of graphite and crystalline hydrates of metals shown in figure 6.

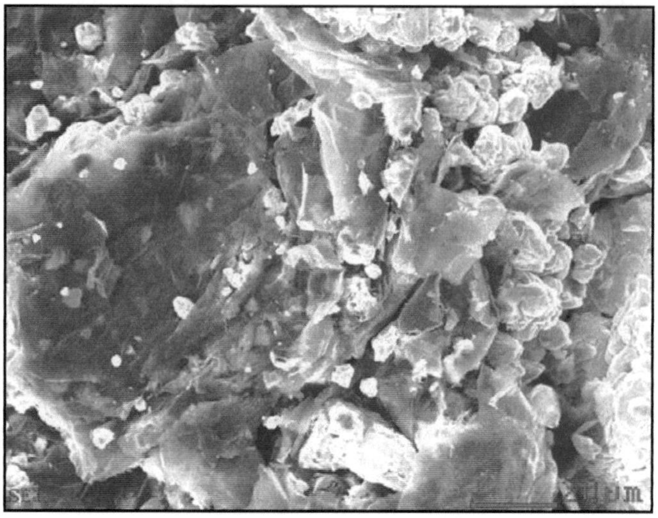

Figure 6. SEM micrograph of expanded graphite.

Figure 6 shows SEM images of the EG samples. The morphology of the EG samples is wormlike and there are a lot of pores that can also be observed on the surface. It is the particular loose and porous structures that would provide EG samples with good adsorption property for the macromolecular compounds.

Figure 7. Oil Sorption capacity of exfoliated graphite.

Figure 7 shows the sorption capacity of expanded graphite. As can be seen from the presented data, the maximum sorption of oil is in the first minutes (~ 3-4 minutes).

4. Conclusions

In this work were to obtain expanded graphite by heat treatment. Morphology of expanded graphite was verified by scanning electron microscopy analysis. Expanded graphite shown outstanding adsorption performance for oil.

References

[1] Kudaybergenov K., Ongarbayev Y., Mansurov Z., Doszhanov Y. Study on the effectiveness of thermally treated rice husks for petroleum adsorption // *Journal of Non-Crystalline Solids* 358 (2012), pp, 2964–69

[2] Ongarbayev Ye.K., Belgibayeva A.S., Kudaybergenov K.K., Mansurov Z.A. Oil Spill Cleanup from Sea Water by Porous Sorbents // *Eurasian Chemico-Technological Journal*, **17**(1), (2015), pp, 41–45

[3] Sorokina H.E., I.V. Nikolskaya, Ionov S.G., Avdeev V. V. Compounds of graphite of acceptor type and new carbon materials on their basis : 2005. T. 54. No. 8.) pp, 1699–1716

[4] Toyoda M., Aizawa J., and Inagaki M. Sorption and recovery of heavy oil by using exfoliated graphite : *Desalination*, 1998. Vol. **4.** №115. pp, 199–201

[5] Sorokina N.N., Avdeev V. V. The composite of nanomaterials on a basis intercalated graphite: 2010. 100 with. to their basis 2005. T. 54. No. 8. pp, 1699–1716

[6] Parvez K., Wu Z.S., Li R., Liu X., Graf R., Feng X., Müllen K. Exfoliation of Graphite into Graphene in Aqueous Solutions of Inorganic Salts : J. Am. Chem. Soc., 2014. Vol. **136**.pp, 6083–91

[7] V. Sridhar, J. Jeon, I. Oh, Synthesis of graphene nano-sheets using eco-friendly chemicals and microwave radiation, *Carbon* **48** (10) (2010)) pp, 1007–09

[8] E.H. Falcao, R.G. Blair, J.J. Mack, I.M. Viculis, C. Kwon, M. Bendicov, Microwave exfoliation of a graphite intercalation compounds, *Carbon* **45** (6) (2007) pp, 1367–69

IOP Conf. Series: Materials Science and Engineering **323** (2018) 012013 doi:10.1088/1757-899X/323/1/012013

The influence of the emulsion composition on the wettability of the emulsion

Yan Jun Liu[1,2,3*], Jian Nan Shao[1,2] and Peng Lei Liu[1,2]

[1]Department of Environmental and Chemical Engineering,Tianjin Polytechnic University,Tianjin 300387, China

[2]Tianjin Engineering Research Center of Textile Fiber Interface Treatment Technology, Tianjin 300270, China

[3]Tianjin Polytechnic University for the textile auxiliaries co., LTD,Tianjin 300270, China

*E-mail: yjliu72@126.com

Abstract. In order to explore the influence of the emulsion composition on the wettability of the emulsion, using lauric acid polyoxyethylene esters (LAE) and polyethylene oleic acid diester (DQA) as the emulsifier and oleic acid ester (QA) as the smoothing agent, the spinning oil emulsion system with the content of smoothing agent above 30% was prepared. The results show that: with the increase of emulsion concentration, the surface tension of emulsion, the contact Angle of emulsion on the surface of the polypropylene fiber and the wetting time of canvas in emulsion all decreases. At the same time,the emulsion has critical micelle concentration, when the concentration is less than CMC, the surface tension of emulsion, the contact Angle of emulsion on the surface of the polypropylene fiber and the wetting time of canvas in the emulsion decreases rapidly with the increase of the emulsion concentration, while it's more than this concentration, the influence of emulsion concentration on the three kinds of nature is smaller. Besides, the increase of the mass fraction of the smoothing agent and the increase of the compound emulsifier HLB will result in worse wettability.

1. Introduction

Wetting is one of the basic properties of microemulsion, during the spinning process, fibers in high-speed drafting, the need for oil quickly and evenly spread to the fiber surface, so as to fully protect the fiber, otherwise, oiling will occur uneven, partially oil-free, large friction, spinnability decreases, so wetting is very important for spinning and processing [1].

The surface tension directly affects the contact angle and the length of the wetting time. Under normal circumstances, the lower the surface tension of the oil is better for wetting permeability of the fiber and oiling effect [2]. There are many ways to determine the surface tension of liquids, such as hanging drop method, deprotection method, bubble pressure method, drop volume method, hanging film method, drop method, capillary rise or fall and so on [3].

After emulsifying the oleate with a single emulsifier,the spin finish has poor effect.The emulsion determined in this paper is mainly composed of the existing surfactants,such as smoothing agents and emulsifiers.The contact angle between the emulsion and the polypropylene fiber was measured by the appearance image analysis method and the wettability was characterized by the hydrophobic fiber [4].

Using the national standard method to initially explore the emulsion concentration and composition of the impact on wettability, in a certain range of spinning oil emulsification can provide some reference.

2. Experimental section

2.1. Experimental materials and instruments

Polyoxyethylene laurate LAE (HLB value = 13.1), polyethylene glycol oleate DQA (HLB value = 10.5), Oleate QA, Tianjin Textile Textile Auxiliary Co., Ltd.

FA2004N electronic balance, Shanghai Jinghai Instrument Co., Ltd. production; JB-2 temperature magnetic stirrer, Shanghai Lei Magnetic New Jing Instrument Co., Ltd. production; DSA25 automatic contact angle measuring instrument, the German company Klv; JYW- 200 automatic surface tension meter, Chengde heyday experimental instrument factory.

2.2. Experimental methods

2.2.1. Emulsion preparation

The emulsifiers LAE and DQA were respectively mixed in different quality proportions of 2: 8, 3: 7, 4: 6 and 5:5 to form a complex emulsifier (represented by C), and the emulsifier C and smooth QA were mixed at different concentrations, of which QA accounted for 30%, 35%, 40% and 60% respectively. Afterwards, water was added and mixed well to prepare emulsions of different concentrations and allowed to stand at room temperature for 24 hours.

2.2.2. Determination of surface tension, contact angle and wettability

The surface tension of the emulsion is measured after being calibrated by absolute ethanol or distilled water at room temperature of 20°C. The densitometer is used to measure the density of the solution and the actual surface tension of the tested oil sample is calculated according to the formula [2].

The contact angles of pure water (concentration 0%) and different concentrations of emulsion on a single polypropylene fiber were measured at 20°C at room temperature using a DSA25 automatic contact angle measuring instrument. Each measurement was averaged five times.

Preparation of a certain concentration of oil emulsion 900mL. Will configure the emulsion at 20°C constant temperature conditions, the use of "surfactant wetting power measurement immersion method" (GB/T 11983-2008) will be submerged in the canvas canvas was measured in the concentration of the emulsion, record Run Wet time, corresponding to the concentration of the emulsion and the ratio of draw wetting curve, can evaluate the impact of the wettability of the emulsion.

3. The results and discussion

3.1. The effect of emulsion concentration on wettability

3.1.1. The contact angle and surface tension

It can be seen from figure 1, compared with pure water,the contact angle of the emulsion on the polypropylene fiber significantly reduced. And under the same emulsion ratio, the contact angle between emulsion and polypropylene fiber decreases with the increase of emulsion concentration. This shows that the experiment of ester-type microemulsion in the fiber on the spread of a very significant increase. This is because emulsion micelles can significantly reduce the surface tension of water as shown in figure 2, making it easier to spread on the hydrophobic fibers. And the higher the concentration of disperse, the smaller the liquid surface tension, the more obvious spread of the emulsion on the fiber.

IOP Conf. Series: Materials Science and Engineering **323** (2018) 012013 doi:10.1088/1757-899X/323/1/012013

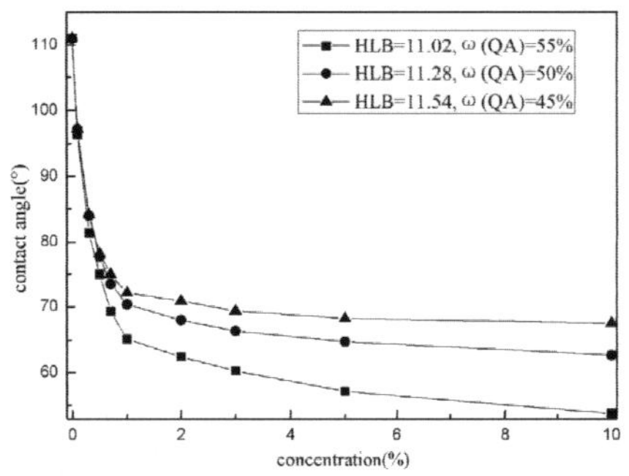

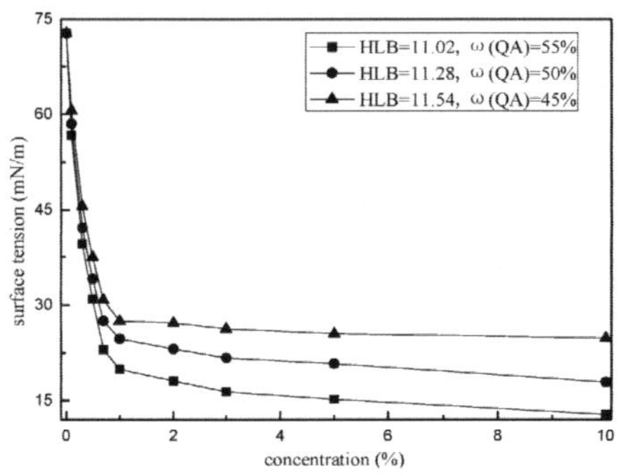

Figure 1. The change of contact Angle of emulsion and fiber along with concentration about three kinds of emulsion with different proportions.

Figure 2. The change of surface tension of emulsion along with concentration about three kinds of emulsion with different proportions.

At the same time, it can also be seen from figure 1 that there is a mutation point (1.0% concentration) in the curve of contact angle with concentration. The concentration at this point can be regarded as the critical micelle concentration (CMC) of the emulsion. The contact angle of the emulsion on the PP fiber decreased rapidly with the increase of the concentration, and the influence of the emulsion concentration on the contact angle decreased gradually after the concentration was increased. This is because the emulsion before reaching the critical micelle concentration by the surfactant in the emulsion, the surface tension decreased rapidly to reach the critical micelle concentration of the emulsion, the concentration continues to increase, the emulsion began to form a large number of plastic oil The degree to which the surface tension of the emulsion decreases will begin to slow as shown in figure 2.

From figure 1 and figure 2, we can see that the contact angle between emulsion and PP fibers is directly related to the surface tension of emulsion, which also confirms the Young's equation: $\gamma^{s} = \gamma^{sl} + \gamma^{l}\cos\theta$, describing the relationship between contact angle and surface tension. From Young's equation, it can be seen that the smaller the surface tension of liquid is, the smaller the contact angle is when the surface tension of the solid is constant.

3.1.2. The wettability

It can be seen from the experiment that the wetting time of the emulsion and the canvas is related to the change of the concentration as shown in figure 3 (a) and figure 3 (b): The wetting time of cloth in pure water is much longer than the wetting time in the emulsion. As the emulsion concentration increases, the wetting time becomes shorter. And when the emulsion concentration is small, as shown in figure 3 (a), the wetting time is rapidly shortened as the emulsion concentration increases. However, when the emulsion concentration exceeds a certain level (about 0.7% in this paper), As shown in figure 3 (b), the wetting time is still shortened as the concentration increases, the wetting time decreases with increasing concentration.

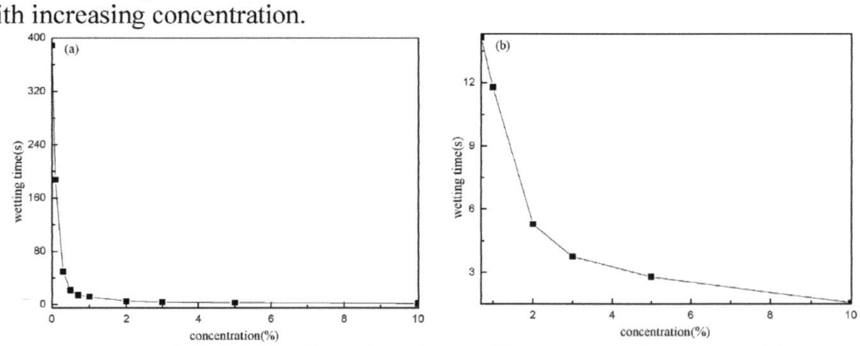

Figure 3. The change of wetting time with concentration of emulsion.

IOP Conf. Series: Materials Science and Engineering **323** (2018) 012013 doi:10.1088/1757-899X/323/1/012013

The reason why the wetting rate of the canvas is much larger than that of the water is because a small amount of spinning oil can obviously reduce the surface tension of the water as shown in figure 3. And when the emulsion concentration is small (less than 0.7%), the surface tension of water decreases rapidly with increasing concentration, while the concentration of more than 0.7%, the rate of surface tension decreases slowly. So there will be a turning point in figure 3 (a) wetting time with concentration changes. It should be noted that, although the rate of slowing, but there are still changes, the same trend. In addition, there is a slight concentration difference between the turning point of the wetting time and the turning point of the surface tension, which may be due to the difference of the solid-phase interfaces measured by the two during the test: The canvas sedimentation method uses canvas when wetting time, its hydrophilic strong, and the measurement of contact angle using a hydrophobic polypropylene fiber, the two interface surface properties vary greatly, resulting in the concentration of these two sets of curves obtained by a certain difference.

3.2. The effect of emulsion composition on contact angle and wettability

3.2.1. The effect of emulsifier ratio on contact angle and wettability

The change of HLB value is caused by the change of the ratio of emulsifier LAE and DQA. The bigger the ratio of LAE to DQA is, the larger the HLB value is. When the QA content is 45% and the emulsion concentration is 1%, the contact angle between emulsion and polypropylene fiber at different HLB and the wetting time of canvas in the emulsion are measured. The results are shown in figure 4. As the HLB value of emulsifier increases Large, contact angle and wetting time are gradually increased, and almost linear relationship.

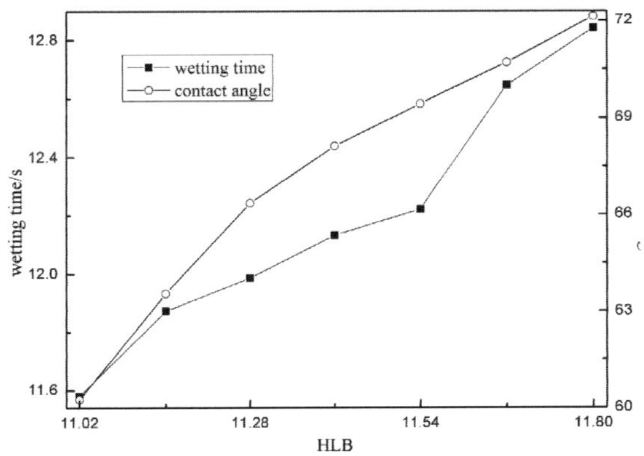

Figure 4. The change of contact Angle and wetting time with the size of HLB under the same ester content (45%).

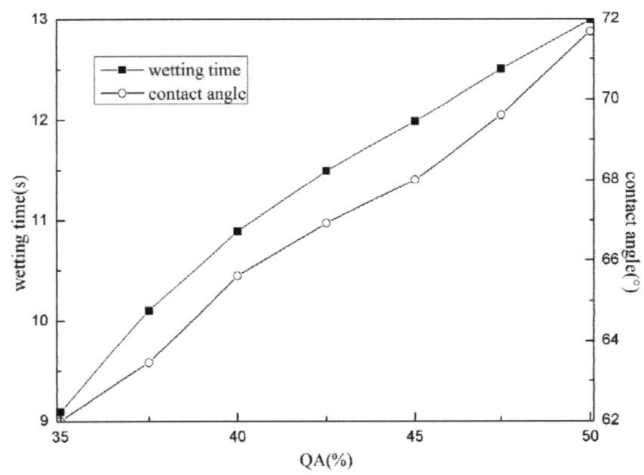

Figure 5. The change of contact Angle and wetting time with the ester content under the same proportion of emulsifier(LAE:DQA =3:7).

Emulsifier can reduce the surface tension between liquid and solid, surface tension drop will lead to the spread of the emulsion on the fiber better, wetting the canvas faster. The single emulsifier lipophilicity and hydrophilic properties have focused. Therefore, the use of a single emulsifier can reduce the surface tension, but its stability is not strong. The composite emulsifier generally contains both hydrophilic and lipophilic emulsifier, it can greatly reduce the surface tension. In order to obtain the best results, we need to mix different emulsifiers, hydrophilic emulsifiers and lipophilic emulsifiers by adjusting the content to balance the hydrophilic emulsifier, lipophilic properties [5].

Different dispersed phases, the continuous phase mutual emulsification, the required HLB values are generally different, and the same dispersed phase, continuous phase required HLB value theoretically should be a fixed value. However, the choice of different emulsifiers for compounding, due to the nature of the different emulsifiers, there may be some deviation [6]. In the experimental range of this paper, the greater the HLB value of the composite emulsifier, the poorer the effect of reducing the surface tension of the emulsion. The larger the contact angle between the emulsion and the polypropylene fiber, the longer the wetting time of the canvas in the emulsion. In addition, the HLB value not only reflects the ratio of emulsifier, but also reflects the size of the hydrophilic base mass of the composite emulsifier. When the hydrophilic mass fraction is larger, the relative lipophilicity will be relatively smaller. Most of the organic synthetic fibers such as polypropylene fibers show hydrophobic-lipophilic properties, so the more hydrophilic hydrophilic emulsion in the fiber the greater the contact angle.

3.2.2. The effect of ester content on contact angle and wettability

When the emulsifier LAE: DQA = 3: 7, the emulsion concentration is 1%, the contact angle between the emulsion and the polypropylene fiber and the wetting time of the canvas in the emulsion under different ester contents were measured. The results are shown in figure 5: The contact angle and wetting time all increased with the increase of ester content, close to the linear relationship.

In a certain concentration and emulsifier ratio, the emulsifier can reduce the liquid surface tension between the liquid [7], the surface tension drops will lead to the spread of the emulsion in the fiber becomes better, but also make the emulsion easier to wet the fiber, canvas gets wet faster in the emulsion. In this experiment, when the emulsion concentration is the same, the greater the QA mass fraction of the lubricant, the lower the concentration of the emulsifier and the lower the surface tension,so the contact angle between the emulsion and the polypropylene fiber and the wetting time of the canvas in the emulsion are therefore greater.

4. Conclusion

Dispersions of the emulsion significantly reduce the surface tension of the emulsion, thus greatly reducing the contact angle between the emulsion and the polypropylene fiber and also greatly enhancing the wettability of the emulsion. And the emulsion surface tension and contact angle, wettability changes with concentration there is a clear turning point, that there is an emulsion critical micelle concentration: less than the critical concentration, the surface tension of the emulsion, the contact angle of the surface of the polypropylene fiber and The wetting time decreased rapidly with the increase of the emulsion concentration, and the effect of the emulsion concentration on the surface tension, the contact angle and the wettability became smaller after exceeding this concentration. At the same smoothing agent level, the greater the HLB value of the emulsifier, the greater the contact angle of the emulsion, the poorer the wettability. At the same emulsifier ratio, the higher the content of the smoothening agent, the larger the contact angle of the emulsion and the poorer the wettability. However, in the range of less than the critical micelle concentration, the influence of these two factors on the wettability is much smaller than the variation of the emulsion's own concentration on the wettability.

Reference

[1] XU Jinyun, WEI Junfu, ZHENG Guo, et al. Study on emulsion property of potassium alkyl-phosphate[J]. Journal of Tianjin Polytechnic University, 2010, 29(2): 45-47.

[2] Han Chunyan, JI Xuan, LI Hongfang. Study on the influence of making up process of chemical fiber preparation agent's surface tension[J]. Synthetic Technology and application. 2014, 29(1): 42-46.

[3] Sripriya R, Muthu R K, Santhosh G, et al. The effect of structure of oil phase, surfactant and cosurfactant on the physicochemical and ebctrochemical properties of biconti-uous microemulsion[J]. Colloid Interface SCI, 2007, 314(2):712-17.

[4] Xie Y, Wang Z, Qin Y, et al. Rheology and wettability of chitosan/hydroxypropyl cellulose blend materials[J]. Journal of Textile Research, 2013, 34(5):25-29.

[5] Shangguan M, Liu Y, Jiao W, et al. Effect of emulsifiers on stability of methanol diesel emulsion fuel[J]. Chemical Industry & Engineering Progress, 2011, 30(3):509-12.

[6] TAN Yuanqiang, HUANG Weijiu, ZHOU Yajun, et al. Study on the relativity between wetting ability and then composition of emulsion[J]. Shanghai Nonferrous Metals, 1999, 20(1): 10-13+46.

[7] Thritima Sritapunya, Boonyarach Kitiyanan, John F. Scamehorn,et al. Wetting of polymer surfaces by aqueous surfactant solutions[J]. Colloids and Surfaces A: Physicochemical and Engineering Aspects, 2012, 409(5): 30-41.

IOP Conf. Series: Materials Science and Engineering **323** (2018) 012014 doi:10.1088/1757-899X/323/1/012014

The study of operating an air conditioning system using Maisotsenko-Cycle

Mohammad S.Khan*, Sami Tahan*, Mohamad Toufic El-Achkar* and Saleh Abou Jamus*

Abu Dhabi University, Abu Dhabi, United Arab Emirates

*E-mail: 1047548@students.adu.ac.ae

Abstract: The project aims to design and build an air conditioning system that runs on the Maisotsenko cycle. The system is required to condition and cool down ambient air for a small residential space with the reduction in the use of electricity and eliminating the use of commercial refrigerants. This project can operate at its optimum performance in remote areas like oil diggers and other projects that run in the desert or any site that would not have a very high relative humidity level. The Maisotsenko cycle is known as the thermodynamic concept that captures energy from the air by using the psychometric renewable energy available in the latent heat in water evaporating in air. The heat and mass exchanger design was based on choosing a material that would-be water resistant and breathable, which was found to be layers of cardboard placed on top of each other and thus creating channels for air to pass through. Aiming for this design eliminates any high power electrical equipment such as compressors, condensers and evaporators that would be used in an AC system with the exception of a 600 W blower and a 10 W fan, thus making it a more environmentally friendly project. Moreover, the project is limited by the ambient temperature and humidity, as the model operates at an optimum when the relative humidity is lower.

1. Introduction

Air conditioners have become a necessity in any household to artificially create and maintain an indoor environment by ventilation, cooling and heating to meet criteria that ASHRAE standards state are within the comfortable ranges. With today's world being focused on preserving energy and making use of wasted energy within a system, renewable energy is considered to be an integral part of any project. Recent studies and researches by Professor Maisotsenko have found an effective and environmentally friendly cycle named as the M-Cycle, which has been used in applications in recent years such as cooling towers, combustion devices and air conditioner [1].

This project aims to design an air conditioning system that uses the M-Cycle in order to cool down and dehumidify ambient air to acceptable standards, with the minimal amount of energy consumption and being environmentally friendly [1-2].

1.1. Impact of the project on the environment

With newer standards looking to completely eliminate all refrigerants including HFCs, this project is independent and is no need of a refrigerant thus eliminating this issue.

As a result, this will reduce the demand for electricity significantly, thus reducing the energy consumed and the amounts of air pollutants and power plants pollution. It can be achieved in terms of using less energy for electrical components that are used in typical AC systems, heat rejection and the use of unwanted refrigerants [3].

The materials used in the project are all recyclable which gives the advantage for the project when it wears out to be recycled and reused again in the same application or another rather than pilling up on the landfills.

1.2. Maisotsenko Cycle and Refrigeration Cycle

Table 1 [4] shows the differences between using an air conditioning system that uses the typical refrigeration cycle against an air conditioning system that operates on a Maisotsenko cycle.

Table 1. Refrigeration cycle and M-Cycle.

Cycles	Advantages	Disadvantages
Refrigeration Cycle	• The refrigerant used is cheap, available and easily. • It is a reversible cycle, in order to turn it to heating.	• Danger of frosting at the valves • The quantity of refrigerant used per ton of refrigeration is high
Maisotsenko Cycle	• Low maintenance cost. • Low pressure drops. • Environmentally friendly cycle with minimal heat rejection.	• Works at lower efficiency when the humidity is higher. • The system cannot be used to cool down a larger residential area.

1.3. Project Objectives

This project aims to reach the following objectives:

- Designing an AC system with low power consumption compared to a vapor compression cycle.
- Designing an AC system with low heat rejection. Designing an AC system which is environmentally friendly.
- Designing an AC system that can meet the standards to provide comfortable conditions in the space.
- Decreasing the maintenance cost as compared to a typical AC system.
- Using the knowledge gained over the past four years and the courses taken during the undergraduate mechanical engineering curriculum [5].

1.4. Outcomes and deliverables

Throughout performing this project, the design process and prototype made, give an expectation of the outcomes that are required for this project:

- Reaching the maximum relative humidity 90-95% for the primary air flow after exiting the heat and mass exchanger.
- Having a comparison between the simulation and prototype results, and attaining the lowest possible percentage error.
- Achieving a temperature drop from the primary air flow inlet to the primary air flow outlet.

2. Literature Review

2.1. Basics of M-cycle

Energy cycles productivity is measured through their efficiency. Dating back to 2003, combined power cycles on average show efficiency of 50%. However, with the introduction and use of the Maisotsenko cycle has increased the efficiency by a further 10%. The cycle operates through 4 main steps where the water and working air enter the heat and mass exchanger [5].

IOP Conf. Series: Materials Science and Engineering **323** (2018) 012014 doi:10.1088/1757-899X/323/1/012014

The water that acts as the working fluid is fractioned off into wet channels in the heat and mass exchanger. The heat is then transferred to the air by evaporation and is rejected. This results in the product fluid exiting with low relative humidity [6].

2.2. Applying M-Cycle to Air Conditioning systems

With air conditioning being a huge necessity for any industry or residential building, a lower energy consumption and more environmentally friendly approach is needed to be taken as electricity plays a huge role in air conditioning for the use of compressors and so forth [6].

The use of indirect evaporative cooling which refers to the Maisotsenko cycle approach can eliminate the fact of increasing the humidity when lowering the air temperature whilst also avoiding any potential health problems due to contaminated droplets that enter into the cooled space [6-7].

2.3. Heat and Mass Exchanger

The main component that allows the Maisotsenko cycle to run with such efficiency is the heat and mass exchanger (HMX). This type of exchanger is formed through several sheets of material known as rigid pad. This material is water resistant and breathable. Several evenly spread pathways made throughout the dry air flow are made, which gradually divert air from the dry channels to the wet channels with an even velocity distribution [7].

Whilst articles and journals have reviewed and focused on heat and mass exchangers, they do not go into the design and set up of the heat and mass exchanger but rather explain how they operate and what it can achieve when being used. This can give a general idea of what a heat and mass exchanger is and how the components of the heat and mass exchanger come together to carry out its objective.

This project aims to further understand and design a heat and mass exchanger by implementing parameters needed to operate it so that it can be practically made in order to use for a Maisotsenko cycle for an air conditioning system. The choice of this material came from experimental work yet the best results were found after trying the cardboard which roughly shows an efficiency of 75-95% [8].

2.4. Mathematical Modelling

2.4.1 for the air stream in the dry channel (secondary air)

Mass balance equation

$$\dot{m}_w = \dot{m}_d + \dot{m}_{w.v} = \dot{m}_d + \dot{m}_d(w) = \dot{m}_d(1+w) \quad (1)$$

$$\dot{m}_d = \dot{m}_D \quad (2)$$

$$w_{in} = w_{out} \quad (3)$$

Energy balance equation

The energy conservation balance considering only sensible heat transfer on the plate surface is given as:

@ Inlet:

$$h_1 = T_{in} + (w_1 * 2501.3) + (1.86 * T_{in}) \quad (4)$$

@ Outlet:

$$\mathrm{h_2 = T_{out} + (w_2 * 2501.3) + (1.86 * T_{out})} \quad (5)$$

Q Lost:

$$Q_{lost} = \dot{m}_D \, \Delta h = \dot{m}_D \, (h_2 - h_1) \quad (6)$$

2.4.2 for the air stream in the wet channel (primary air)

Mass Balance Equation

The mass conservation balance for the water vapor inside the wet channel:

@ Bottom channel

$$\dot{m}_w = X\dot{m}_w + (1 - X)\dot{m}_w \quad (7)$$

@ Upper channel:

$$m^{\cdot}(1-nX)+Xm^{\cdot}=m^{\cdot}(1-X(n-1)) \quad (8)$$

where:
X: Represents the fragment of the air that travels through the material to the upper channel.
n: Represents the number of needed sections in one channel
Energy balance equation
@ Inlet:

$$t_1=t_{in} \quad and \quad w_1=w_{in} \quad (9)$$

$$h_1=0.24T_1+(w_1)*(1061.2+0.444T_1) + (1.86*T_{in}) \quad [9] \quad (10)$$

where:

$$h_x=h_1 \quad (11)$$

@ Outlet:

$$m^{\cdot}_w(1-nX)(h_1)+(m^{\cdot}_w X)(h_1)+m^{\cdot}_v h_w = m^{\cdot}_w(1-X(n-1))h_2 \quad (12)$$

2.4.3 Heat Exchanger

$$Q=\frac{T_1-T_2}{\sum R} \quad (13)$$

3. Prototype

Several iterations and designs are made until the final design in terms of material and the geometry of the model itself. Building the prototype consists of building a heat and mass exchanger with a material that fits the specifications of being breathable and water resistant. This heat and mass exchanger is placed within a box made of a thermally insulating material and therefore having minimal heat rejection. A heat exchanger is placed on the top of the box to allow heat exchange between the primary and secondary air streams. The aim of this process is to lower the dry bulb temperature.

3.1. Material Selection

The material selection process mainly requires finding a material fitting the specifications of being breathable and water resistant in order to fulfill the requirements needed for an indirect evaporative cooling process. Materials such as polypropylene and fiber mesh were tested.

In the last iteration, the decision was reached to use a heat and mass exchanger made out of a material named rigid pad which had properties that matched the specifications that were set. Several layers were placed on top of each other with curvatures creating numerous channels for the primary air to pass through them. The channels are made in several opposing directions and hence giving a greater area for the air to disperse in.

3.2. Prototype Setup

The heat and mass exchanger used is of certain dimensions 0.45×0.08×0.13 in meters, a distance of 0.05m is left between the heat and mass exchanger and the acrylic glass box, in order to have the remaining air flow pass over it and collect the air that has been cooled down and humidified. With a fiber mesh material attached at the bottom of the heat and mass exchanger in order to hold any excess water that would otherwise drip onto the bottom of the acrylic housing.

With the heat and mass exchanger being setup as explained, the rest of the prototype design can be completed. The assembly will show that within the acrylic glass box, a plastic tube will be placed on top of the heat and mass exchanger where it will be fed with water and will act as a sprinkler.

The second part of the prototype consists of a simple heat exchanger, where the primary air flow that has passed through the heat and mass exchanger will exchange heat with a secondary air flow that is simply ambient air which would be passed in a pipe above it, as they would be separated by an aluminum sheet. This setup is displayed in figure 1.

4. Results and Discussion

4.1. Simulation Results

After running the C++ code and EES program, the results obtained for the temperatures in graph as shown in figure 2. Where at 15% humidity level the temperature has dropped from 27.0°C to 15.2°C. Whilst under high humidity levels such as 80% the drop decreases to about a difference of 1.9°C. The results show what was expected and that is that the optimum performance for this design is when the humidity is at the minimal and the ambient temperature is at a maximum, this showed a temperature drop of 18.2°C.

These results have proved that using the water as the working fluid have affected the temperature drop that may not be similar to that of using typical refrigerants such as R134-a but are still within what is considered to be satisfying results. Moreover, the advantage to using the water over a refrigerant is the ease of availability and being more environmentally friendly.

Figure 2 shows 4 different curves for the simulation results at different humidity levels. This proves the analysis made before running the simulation that under higher humidity levels the temperature drop will decrease and not reach comfortable levels. This is due to the fact that the primary air inlet has a higher relative humidity when entering into the heat and mass exchanger and thus has a less of a capacity to absorb moisture and cool down.

Figure 3 and figure 4 shows the simulation result with different flow rate and cross section area respectively. As shown in figure 3 the drop in temperature increases when the flow rate increased because as the flow rate increase there will be more air molecules which can evaporate more water causes the air to be colder.

Where figure 4 shows that as the cross section increases the drop in temperature increases too because the heat transfer in function of the surface area which means as the cross section area increases there will be more heat transfer causes the air to be colder.

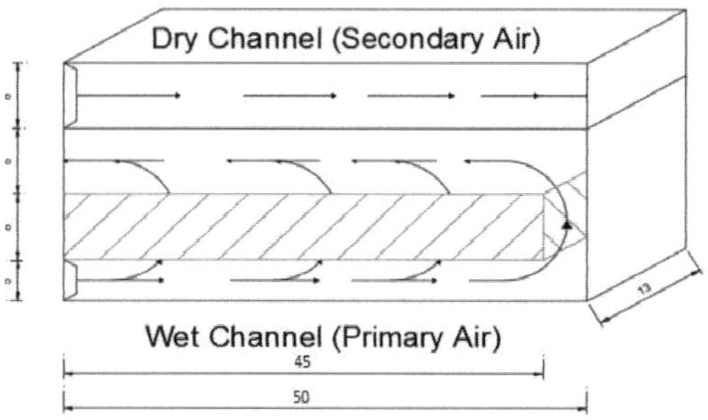

Figure 1. Prototype setup.

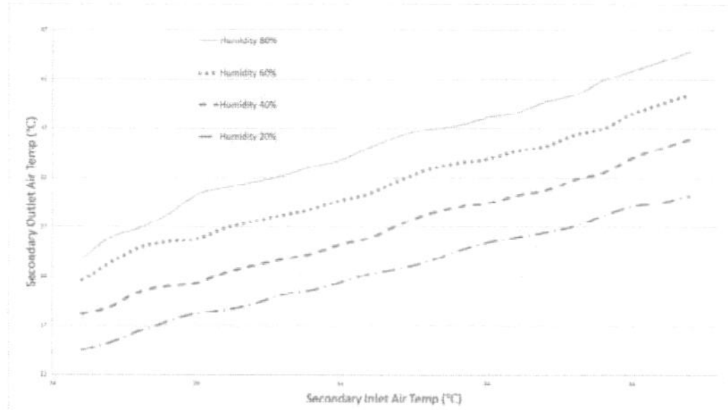

Figure 2. Simulation Results with different relevant humidity.

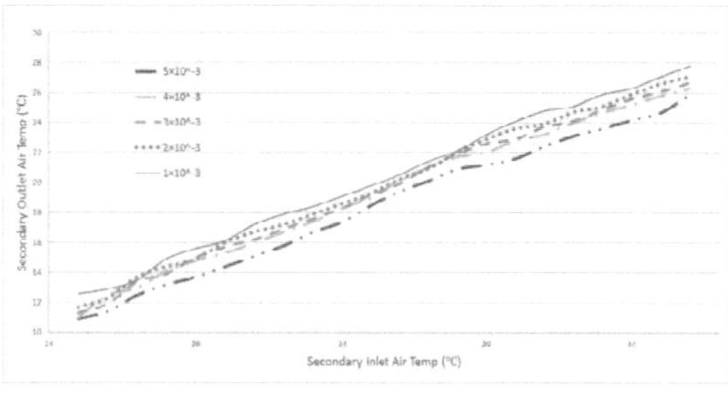

Figure 3. Simulation result with different flow rate.

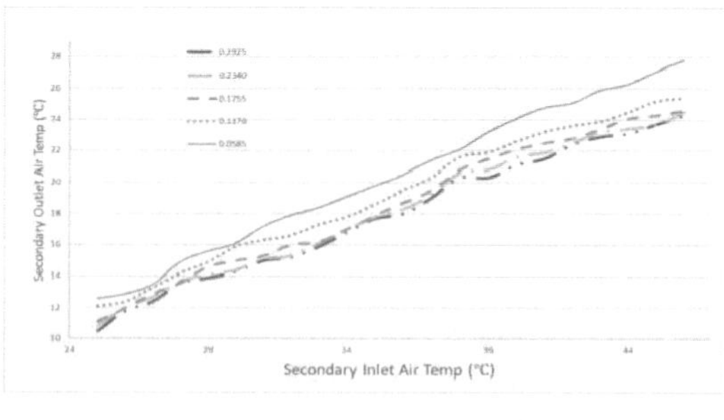

Figure 4. Simulation results with different cross section area.

4.2. Prototype Results

Figure 5 the difference noticed is caused by a major factor other than how many loops the air went through in the heat and mass exchanger which is the humidity level of the primary air. The more humid the air is, the less cool air provided from the secondary air stream. The rises and falls are caused by the difference in the humidity levels of the air provided, where at the low humidity it is noticed that the secondary air is cooler.

After the results had been obtained, figure 6 shows the experimental results with those found through the simulation. The graph shows similar results with the maximum difference around 1°C at the 25°C. Percentage error was calculated for the entire values and showed an average of 5.1%. The difference found between the simulation and prototype results are due to the reasons that the primary inlet air flowing into the heat and mass exchanger had not completely passed through it and some of it had leaked from either side of the housing as fixing the heat and mass exchanger was not completely secured.

Another reason that could've shown this error is that the housing used for the model does not have perfect insulation and will have heat escaping. With the values being obtained from the simulation and experiments, the coefficient of performance can be found to be according to these calculations as:

$$\frac{Cooling\ load}{W_{in}} = \frac{6307.2\ W}{600\ W} = 10.512$$

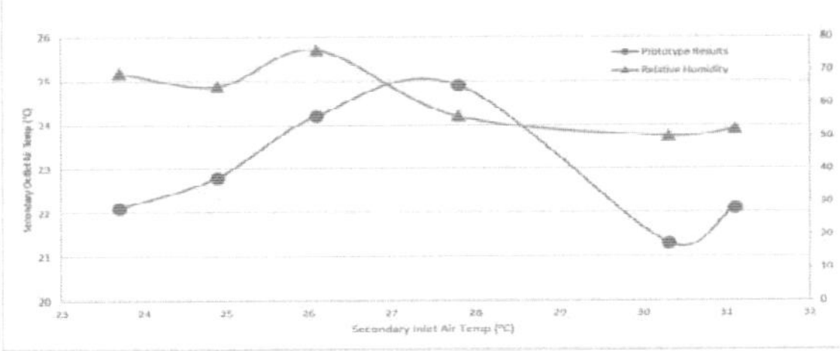

Figure 5. Prototype results.

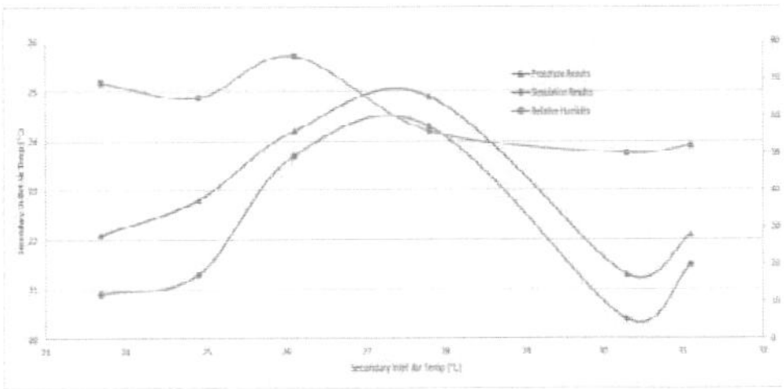

Figure 6. Prototype vs simulation.

5. Future Developments

Due to some of the restrictions that were faced throughout this project such as the budget and timing limitations, not all ideas and objectives could be completed for the project. Therefore, some of the future developments that can be included for the project are as follows:

As the design currently runs on a minimal electrical energy, it is possible to run the system on completely green energy and be independent of electricity. This can be done by powering the blower and fan with solar panels.

The current model designed can only output a certain temperature as according to the tabulated results, which gives a limitation to the output temperature and cannot be set by the user as desired. This can be improved by having the consumer choose the desired temperature within the available values. This can be done by means of controlling the mass flow rate of the primary air flow. The setup for this development is done by adding an electronic damper at the tip of the blower.

As the humidity was one of the limitations of the system which would limit the temperature drop that would occur for the secondary air, integrated liquid desiccant can be utilized. This cycle can be placed before the primary air inlet in order to decrease the humidity of the outside air, considering humidity levels are relatively high in the Middle East region. As the humidity drops this allows for a more efficient cycle.

6. Conclusion

The main objective of this project was to design an air conditioning system that would have a minimal electrical energy consumption and for it to be environmentally friendly. This would mean designing a system that would be independent of the common electrical components in a typical air conditioning unit including a compressor, evaporator and condenser.

Such a design relieves the user of much of the maintenance issues due to lack of electrical components. Moreover, the absence of a compressor would eliminate the use of a refrigerant and thus making the system more environmentally friendly.

To conclude, running the prototype and displaying the results showed that the main objective of using a green and lower power consumption air conditioning design operated on a Maitsotsenko cycle was achieved by reaching a cooling load of 2.0 TR under optimum conditions. However, due to the system being a function of relative humidity which is an uncontrollable variable under simulation results but only reached 1.87 TR due to the weather conditions under the prototype run. Achieving this was in terms of, increasing the humidity and dropping the temperature of the primary air stream, in order to heat exchange with the secondary air stream.

Acknowledgement

Firstly, all thanks to God and our families for their support and guidance throughout the capstone project. We are thankful for our advisor Dr. Mohammad Sultan Khan for his guidance and experience to help us complete this capstone report. We would like to extend our appreciation to Dr. Hamdi for his help in figuring out how to deal with experimentation to find out the best suitable material that fits our specifications. We are grateful towards Engineer Nasr Al Khudari for his help in MATLAB and EES guidance, which were critical in our project and Engineer Abed Abdelaziz for his tips on how to write the report up to the professional standards.

References

[1] P. a. D. A. K. Glanville, "Technology Fundamentals for the Dew Point Evaporative Cooling (The Maisotsenko-Cycle) Concept," GTI (2015).
[2] S. Anisimov, D. Pandelidis and M. V,"NUMERICAL ANALYSIS OF HEAT AND MASS TRANSFER PROCESSES THROUGH THE MAISOTSENKO CYCLE," Institute of Air Conditioning and District Heating, (2014).
[3] S. M. Anisimov, D. Pandelidis, A. Jedlikowski and V. Polushkin, "Performance investigation of a M (Maisotsenko)-cycle cross-flow heatexchanger used for indirect evaporative cooling," elsevier, (2016).
[4] Amer, "A Review of Evaporative Cooling Technologies," International Journal of Environmental Science and Development, vol. **2**, no. 6, (2015).
[5] D. Pandelidis, S. Anisimov and W. M. Worek, "Performance study of the Maisotsenko Cycle heat exchangers in different," International Journal of Heat and Mass Transfer, (2014).
[6] Chandrakant Wani, "A Review on Potential of Maisotsenko Cycle in Energy," International Journal of Advance Research in Science, vol. **01**, no. 01, (2012).
[7] T. D. Rogdakis ED, "Maisotsenko cycle: technology overview and energy-saving potential in cooling systems," vol. **3**, (2015).
[8] K. Buscemi, "Refrigeration & Air Conditioning: Converting Disbelievers," 1 4 (2005).
[9] F. C. McQuiston, J. D. Parker and J. D. Spitler, Heating, Ventilating and Air Conditioning Analysis and Design, WILEY, (2011).

IOP Conf. Series: Materials Science and Engineering **323** (2018) 012015 doi:10.1088/1757-899X/323/1/012015

On justification of efficient Energy-Force parameters of Hydraulic-excavator main mechanisms

Anatoliy Komissarov[1,2], Yuliya Lagunova[1,2], Viktor Shestakov[1] and Olga Lukashuk[2]

[1]Ural State Mining University, Yekaterinburg, Russia
[2]Ural Federal University, Yekaterinburg, Russia

*E-mail: ptmir@inbox.ru

Abstract. The article formulates requirements for energy-efficient designs of the operational equipment of a hydraulic excavator (its boom, stick and bucket) and defines, for a mechanism of that equipment, a new term "performance characteristic". The drives of main rotation mechanisms of the equipment are realized by hydraulic actuators (hydraulic cylinders) and transmission (leverage) mechanisms, with the actuators (the cylinders themselves, their pistons and piston rods) also acting as links of the leverage. Those drives are characterized by the complexity of translating mechanical-energy parameters of the actuators into energy parameters of the driven links (a boom, a stick and a bucket). Relations between those parameters depend as much on the types of mechanical characteristics of the hydraulic actuators as on the types of structural schematics of the transmission mechanisms. To assess how energy-force parameters of the driven links change when a typical operation is performed, it was proposed to calculate performance characteristics of the main mechanisms as represented by a set of values of transfer functions, i.e. by functional dependences between driven links and driving links (actuators). Another term "ideal performance characteristic" of a mechanism was introduced. Based on operation-emulating models for the main mechanisms of hydraulic excavators, analytical expressions were derived to calculate kinematic and force transfer functions of the main mechanisms.

1. Introduction

Given the current state of competitive development of mining industry, it is no wonder that the problem of increasing manufacturing machinery quality (in particular, operational-equipment energy efficiency) is becoming a top-priority issue.

The scientific basis and main principles behind efficient operation of hydraulic excavators (as the main type of excavation machinery) were developed and formulated in the papers authored by Melnikov N. N., Ranev A. V., Satovskiy B. I., Steinzeig V. M. and other scientists. Further development of the theory of hydraulic excavators could be found in [1-7].

In a general case, a preliminary estimate of the equipment energy efficiency could be provided by means of comparing capacity-specific energy consumption to theoretical energy intensity of the process in question.

But it does not allow to evaluate the rate of energy consumption without assessing formation dynamics for the energy-force parameters realized on an operating element – the performance characteristic of a mechanism determining its operating potential.

2. The goal of the research and its tasks

The goal of the research is to increase the operational energy efficiency of hydraulic excavators by defining the ideal performance characteristics of their main mechanisms.

The tasks which this paper is due to accomplish are:

- to assess formation dynamics for the energy-force parameters realized on the links driven by the main mechanisms (a boom, a stick and a bucket);
- to determine performance characteristics of the main mechanisms;
- to define ideal performance characteristics of the mechanisms.

3. The solution of the research tasks

The object of the research are the main mechanisms of the operational equipment of a hydraulic excavator (the rotation mechanisms of its boom, stick and bucket). The diagrams for determining the coordinates of design-diagram points are shown in figures 1, 2, 3, and the ones for evaluating transfer functions – in figures 4, 5, 6.

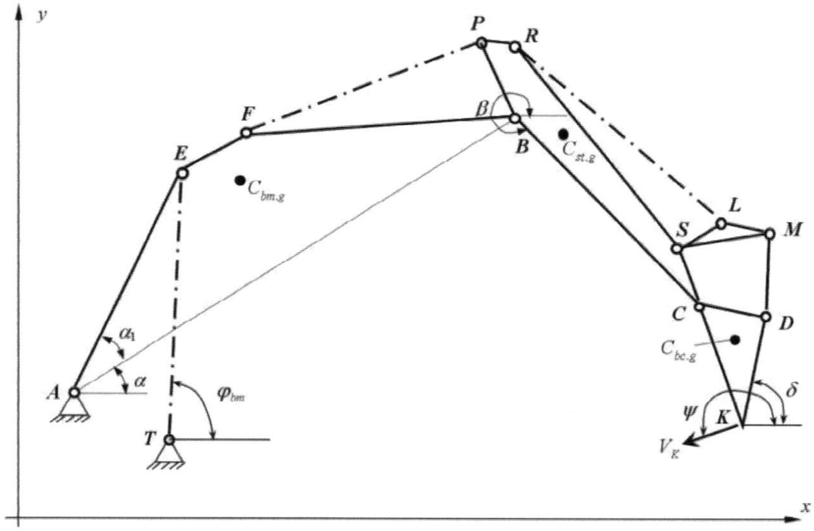

Figure 1. Schematic of the operational equipment.

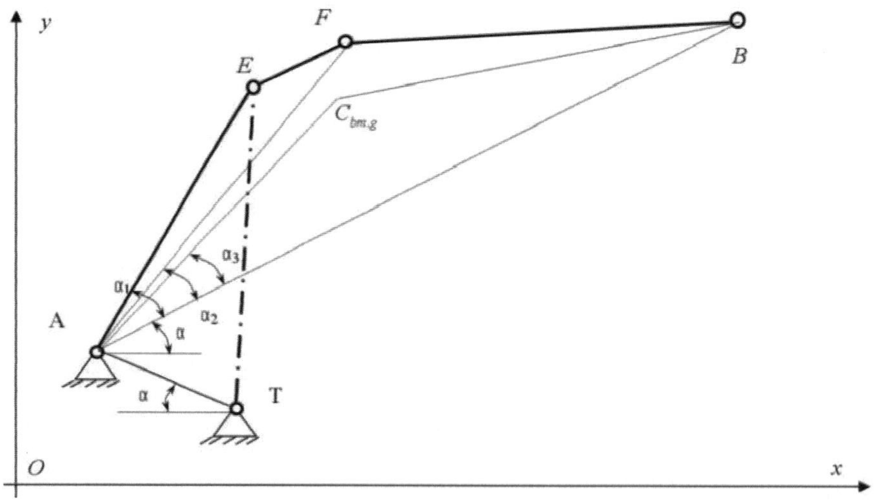

Figure 2. Diagram for determining coordinates of the boom elements.

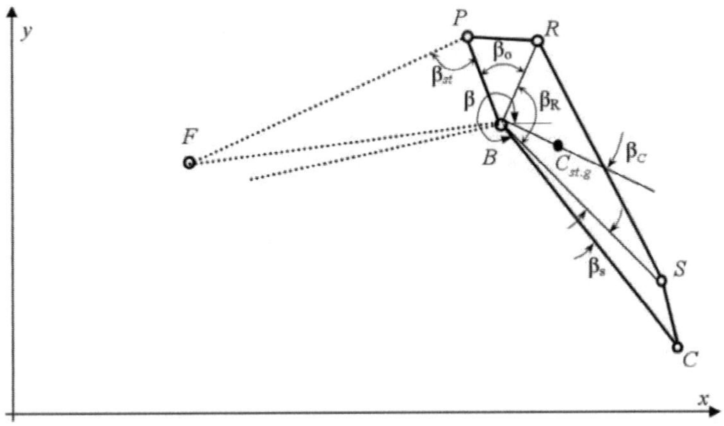

Figure 3. Diagram for determining parameters of the stick.

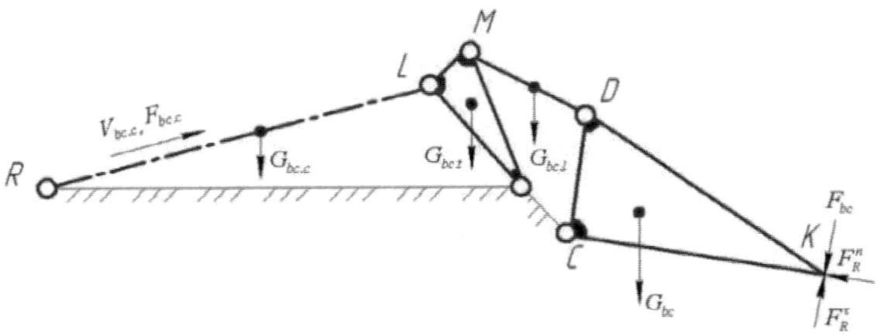

Figure 4. Diagram for determining transfer functions of the bucket-rotation mechanism: RL – length of the hydraulic cylinder (the generalized coordinate); $V_{bc.c}$ – the speed of the cylinder rod; $F_{bc.c}$ – the force on the cylinder rod; F_R^n and F_R^τ – the normal and tangential forces of excavation resistance; F_{bc} – the force on the cutting edge (teeth) of the bucket; $G_{bc.c}$, $G_{bc.t}$, $G_{bc.l}$, G_{bc} – the gravity forces for the bucket cylinder, actuator (tipping) link, bucket link and bucket.

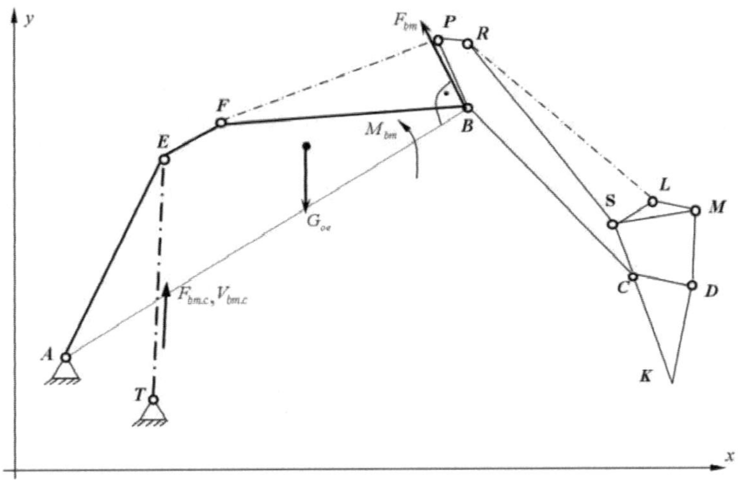

Figure 5. Diagram for determining transfer functions of the boom-rotation mechanism: $V_{bm.c}$ – the speed of the cylinder rod; $F_{bm.c}$ – the force on the cylinder rod; F_{bm} – the force on point B, which

generates M_{bm} – the torque relative to point *A* and equal to the moment of the force $F_{bm.c}$; G_{oe} – the gravity force for the operational equipment

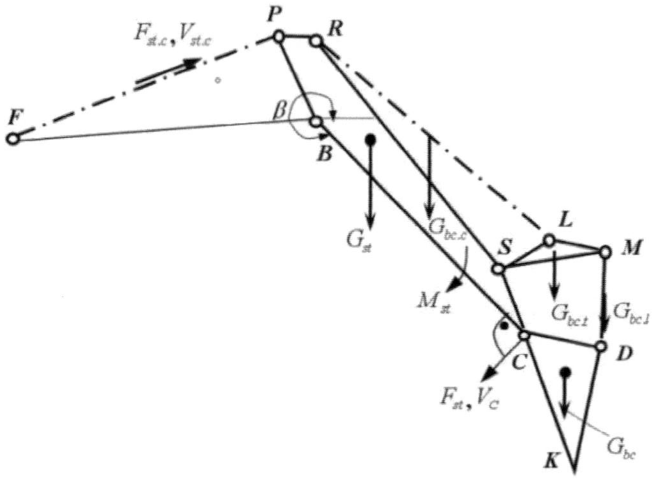

Figure 6. Diagram for determining transfer functions of the stick-rotation mechanism: $V_{st.c}$ – the speed of the cylinder rod V_C; – the speed of point *C*; $F_{st.c}$ – the force on the cylinder rod; F_{st} – the force on point *C*, which generates M_{st} – the torque relative to point *B* and equal to the moment of the force $F_{st.c}$; G_{st}, $G_{bc.c}$, $G_{bc.t}$, $G_{bc.l}$, G_{bc} – the gravity forces for the stick, bucket cylinder, actuator (tipping) link, bucket link and bucket.

One feature of the main mechanisms of hydraulic operational equipment is kinematic linkage between hydraulic actuators (hydraulic cylinders) and driven links of the mechanisms. The linkage exists because the actuators themselves (or the cylinders, their pistons and piston rods) act as links of transmission mechanisms.

At the same time, relations between mechanical-energy parameters of the hydraulic actuators and energy-force parameters of the driven links (a boom, a stick and a bucket) depend both on the types of mechanical characteristics of the actuators and on the types of structural schematics of the transmission (leverage) mechanisms.

Based on operation-emulating models for the main mechanisms of hydraulic-excavator equipment [8, 9], analytical expressions were derived to calculate kinematic H_V and force H_F transfer functions of the main mechanisms, i.e. functional dependences between driven links and driving links (actuators).

The main excavating mechanism is the bucket-rotation mechanism. Its transfer functions are defined as:

$$H_{V.bc} = \frac{V_K}{V_{bc.c}} = \frac{CK \cdot MS \sin(\angle DMS)}{CD \cdot RS \sin(\angle LRS) \sin(\angle CDM)}; \quad (1)$$

$$H_{F.bc} = \frac{F_{bc}}{F_{bc.c}} = \frac{CD \cdot RS \sin(\angle LRS) \sin(\angle CDM)}{CK \cdot MS \sin(\angle DMS)} + \Delta; \quad (2)$$

$$\Delta = \frac{CD \cdot \sin(\angle CDM)}{F_{bc.c} CK} \left[\frac{G_{bc}(X_{bc.g} - X_C) - 0{,}5G_{bc.l}(X_C - X_D)}{CD \sin(\angle CDM)} - \frac{G_{bc.t}(X_S - X_{bc.t.g}) + 0{,}5G_{bc.c}(X_S - X_L) + 0{,}5G_{bc.l}(X_S - X_M)}{MS \sin(\angle DMS)} \right],$$

where V_K – the speed of excavation (for point *K*); $V_{bc.c}$ – the speed of the cylinder rod; F_{bc} – the force on the cutting edge (teeth) of the bucket; $F_{bc.c}$ – the force on the cylinder rod;

$\angle DMS, \angle LRS, \angle CDM$ – angles; G_{bc}, $G_{bc.l}$, $G_{bc.t}$, $G_{bc.c}$ – the gravity forces for the bucket, the bucket link, the actuator (tipping) link and the bucket cylinder; X_C, X_D, X_S, X_L, X_M – the coordinates of points; $X_{bc.g}$, $X_{bc.t.g}$ – the center-of-gravity coordinates for the bucket and the actuator (tipping) link.

The positions of the bucket-rotation mechanism links and the angles $\angle DMS$, $\angle LRS$, $\angle CDM$ are determined by the generalized coordinate RL and the fixed parameters of the mechanism – the linear and angular parameters of the links.

The point coordinates for the mechanism X_R, X_S, ..., X_D and the center-of-gravity coordinates for the bucket and the actuator link $X_{bc.g}$, $X_{bc.t.g}$ are determined in the Cartesian coordinate system with its center set at the point of intersection between the rotation axis of the rotating platform (axis *Y*) and the datum level of the excavator (axis *X*) (figure 1).

The value of Δ is determined by the ratio between the gravity force of the equipment elements and the force on the cylinder rod $F_{bc.c}$, but actually it also depends on the bucket position.

Δ is at its maximum (for the bucket (line *CK*) positioned horizontally) in two cases: in the beginning of excavation (when the forces G_{bc} and F_{bc} act in the same direction), with $\Delta_{beg.max} = 0,02$; and in the end of excavation (when the forces G_{bc} and F_{bc} act in the opposite directions), with $\Delta_{end.max} = -0,06$.

On the whole, the position of the bucket also determines its load duty. In the beginning (when the force G_{bc} and the tangential force of excavation resistance F_R^{τ} act in the opposite directions), the summation of external loads is at its minimum; in the end of excavation, it increases.

Table 1 cites the values of the transfer functions calculated for the bucket-rotation mechanism of a EGO-150 excavator by Uralmashplant (while the excavator in the process of excavation is extending the bucket cylinder rod [10]).

Table1. Calculation results of bucket-rotation mechanism transfer functions.

Transfer functions	Bucket-cylinder length, m				
	3,23	3,645	4,06	4,475	4,89
$H_{V.bc}$	4,82	2,64	2,49	2,75	3,69
$H_{F.bc}$ *	0,206	0,378	0,402	0,366	0,270

The relative values of the force parameters (and accordingly, the forces on the bucket teeth) are increasing until the cylinder rod has been extended to an approximately one-half length of the rod stroke, and then they are decreasing. This means that the performance characteristic of that bucket-rotation mechanism does not provide correspondence between the values of the force parameters realized on the bucket and the bucket load duty (characterized by increase in external-load rate during excavation). It follows that, during soil excavation, the hydraulic actuator becomes overloaded by the end of the stroke, which could trigger its safety valve and stop the whole process.

Thus, the nominal mode of operation for the hydraulic actuator of the bucket-rotation mechanism (with the installed capacity of the actuator being used to its full) is realized (in the case of the ideal performance characteristic of the bucket-rotation mechanism) in the form of a monotonically increasing function.

For the boom-rotation mechanism, the transfer functions will be (figure 2):

$$H_{V.bm} = \frac{V_B}{V_{bm.c}} = \frac{AB \cdot ET}{AE \cdot AT \sin x}; \quad H_{F.bm} = \frac{F_{bm}}{F_{bm.c}} = \frac{AE \cdot AT \sin x}{AB \cdot ET}; \tag{3}$$

$$x = \arccos\left(\frac{AE^2 + AT^2 - ET^2}{2AE \cdot AT}\right),$$

where $V_{bm.c}$ – the speed of the cylinder rod; V_B – the speed of point *B*; $F_{bm.c}$ – the force on the cylinder rod; F_{bm} – the force on point *B*, which generates M_{bm} – the torque relative to point *A* and equal to the moment of the force $F_{bm.c}$.

For the stick-rotation mechanism, the transfer functions will be (figure 3, 6):

$$H_{V.st} = \frac{V_C}{V_{st.c}} = \frac{BC}{BF \sin y}; \tag{4}$$

$$H_{F.st} = \frac{F_{st}}{F_{st.c}} = \frac{BF \sin y}{BC}; \tag{5}$$

$$y = \arccos\left(\frac{BF^2 + FP^2 - BP^2}{2BF \cdot FP}\right),$$

where $V_{st.c}$ – the speed of the cylinder rod; V_C – the speed of point *C*; $F_{st.c}$ – the force on the cylinder rod; F_{st} – the force on point *C*, which generates M_{st} – the torque relative to point *B* and equal to the moment of the force $F_{st.c}$.

The values of the force transfer functions of the boom-rotation and bucket-rotation mechanisms have decreased by the end of their cylinder rod strokes. This means that, at that moment, the hydraulic actuators of the mechanisms become overloaded, which could trigger their safety valves.

4. Conclusion

The analysis of the performance characteristics of operational-equipment mechanisms allows to assess the impact which the kinematic schemes of the mechanisms have on the operational efficiency of those mechanisms and a hydraulic excavator as a whole.

The synthesis of the ideal performance characteristics of operational-equipment mechanisms allows to increase usage of the installed capacity for the driving actuator and, ultimately, provide higher values of engineering-and-economical operational characteristics for hydraulic excavators.

References

[1] Borshh-Komponiec L. V. Methodology of prompt assessment for open-pit hydraulic excavators // Mining industry. 1996. № **1**. pp. 29-37.
[2] Komissarov A. P., Lautenshlejger A. A., Suslov N. M. Assessment of energy parameters of hydraulic-excavator operational equipment // Heavy engineering industry. 1991. № **8**. pp. 25-29.
[3] Bules P. Operating efficiency of open-pit excavators with electromechanical and hydraulic drives for main mechanisms // Mining industry. 2014. № **6** (118). pp. 36-37.
[4] Visbek Z. et al. On efficiency of using open-pit hydraulic excavators // Mining industry. 1998. № **5**. pp. 25-29.
[5] Dudczak A. Excavators: theory and design. Warsaw: PWN, 2000.
[6] Hall A. Characterizing the operation of a large hydraulic excavator. Master Diss. School of Engineering the University of Queensland, Brisbane, Australia, 2002. 150 p.
[7] Park B. Development of a virtual reality excavator simulator: a mathematical model of excavator digging and a calculation methodology. PhD Diss. Virginia Polytechnic Institute and State University. Blackburg, Virginia, USA, 2002. 223 p.
[8] Komissarov A. P., Shestakov V. S. Operation-emulating model for hydraulic-excavator operational equipment // Mining equipment and electromechanics. 2013. № **8**. pp. 20-24.
[9] Komissarov A. P., Lagunova Ju. A., Shestakov V. S. Interconnections between design and operation parameters of hydraulic operational equipment of excavators // Mining equipment and electromechanics. 2014. № **11**. pp. 9-14.
[10] Pobegajlo P. A. Heavy-duty single-bucket hydraulic excavators. Selection of main geometric parameters of operational equipment at early stages of designing. M.: Lenand, 2014. 296 p.

IOP Conf. Series: Materials Science and Engineering **323** (2018) 012016 doi:10.1088/1757-899X/323/1/012016

Operation of single-bucket excavator transmission system

Olga Lukashuk[1], A P Komissarov[1,2] and K Yu Letnev[1]

[1]Ural Federal University, 620002, 19 Mira Street, Yekaterinburg, Russia
[2]Ural State Mining University, 620014, 85 Khokhryakova Street, Yekaterinburg, Russia

*E-mail: oldim96@mail.ru

Abstract. The goal of the research was to determine the mechanism behind formation of operation parameters for the main mechanisms of a single-bucket excavator. It was established that when the main mechanisms of the excavator (hoisting and thrusting mechanisms) and the mechanism of its operational equipment (all of them being parts of its transmission system) act jointly, the modes of hoisting and thrust mechanisms operation change significantly. Operation parameters of the excavator's main mechanisms (hoisting and thrusting velocity and forces) were evaluated from the positions of its bucket in the work area.

1. Introduction

For a single-bucket excavator as a complex electromechanical system, the efficiency of its operation are determined by the degree of correspondence between operation parameters of its main mechanisms (hoisting and thrusting mechanisms) and coordination between them during the operation which is carried out by the excavator's end effector in accordance with the law of its motion.

The scientific basis and main principles behind efficient operation of hydraulic excavators (as the main type of excavation machinery) were developed and formulated in the papers authored by Melnikov N. N., Ranev A. V., Steinzeig V. M. and other scientists. Further development of the theory of hydraulic excavators could be found in [1-8].

The law of bucket motion (path, rock resistance to excavation) depends on mining and technical conditions of operation (rock density, presence of oversize blocks, height of a working face slope and so on). A change in the conditions leads to a significant scatter in the values of operation parameters and hinders the process of rock excavation.

Determining the mechanism behind formation of those values for various conditions of operation would allow to validate the selection of an efficient control algorithm for the working procedure and use full capacity of a single-bucket excavator's power unit.

2. The Solution of the Tasks

The goal of the research is to determine the mechanism behind formation of operation parameters for the main mechanisms of a single-bucket excavator (that is the capacity used in the process of excavation).

The tasks of the research are:

- to evaluate the operation parameters of the main mechanisms from the positions of the excavator's bucket in the work area;
- to determine the degree of interaction between the main mechanisms in their joint action.

3. The Goal of the Research and its Tasks

One special feature of the electromechanical system of a single-bucket excavator (figure 1) is its two-stage transmission gear, which consists of the excavator's main mechanisms (hoisting and thrusting mechanisms) and the mechanism of its operational equipment (connecting the main mechanisms and its end effector (the bucket)).

The operational equipment mechanism (figure 2) includes four movable links: link *1* in the form of rigidly connected stick *AB* and bucket *BK* ("stick-bucket link"), which forms a two-degrees-of-freedom kinematic pair (translational and rotational freedoms) with stand O_1 via a saddle bearing – link *2*; link *3* – an element of the hoisting rope and link *4* – lead block of the boom. Links *3* and *4* are kinematically equivalent to the rods which form rotational pairs with each other, link *1* and stand O_2. Links *2* and *4* are output links of the thrusting and hoisting mechanisms. And then, the velocity of point *A*, which is the same for links *1* and *2*, is equal to the velocity of thrust V_T, and the velocity of point *D* – equal to the velocity of hoisting V_H.

That mechanism has two degrees of freedom, that is the number of possible independent motions (generalized coordinates) equals two. The coordinates could be established in the positions of the links *2* and *4* or – as the link *1* forms a two-degrees-of-freedom kinematic pair with the stand – in the position of the link *1*. Since the positions of the links *2* and *4* are arbitrary due to uncoordinated functioning of the hoisting and thrusting mechanisms, then the coordinates of point *K* (the end of the link *1* and the top of the cutting edge of the bucket or its teeth) in the fixed coordinate system *X-Y* and the inclination of the point *K* path of motion (direction of the excavation velocity) are taken as the generalized coordinates.

The fixed coordinate system consists of two mutually perpendicular axes which intersect in point *O* (*OX* – the datum level of the excavator, *OY* – the rotation axis of its platform). Since the position of the bucket determines the positions of all links of its mechanisms, then the link *1* could be taken as the initial link. Which means, to determine the positions and velocities of the links in the mechanism of operational equipment and the values of operation parameters for the main mechanisms, one should start by determining (setting) the positions and velocity of the link *1* (the top of the cutting edge).

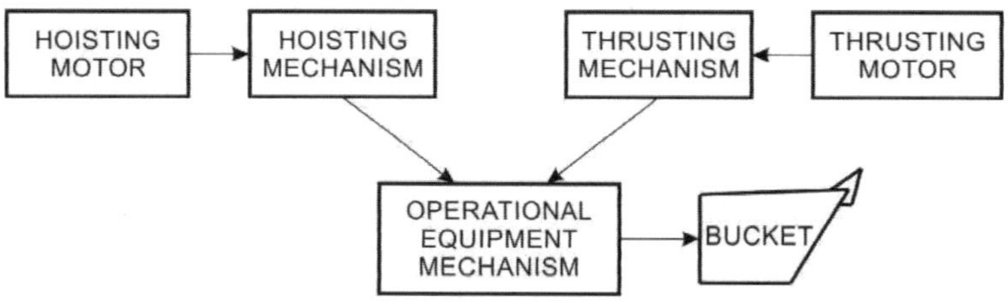

Figure 1. Structural diagram for the electromechanical system (excavation mechanisms) of a single-bucket excavator.

On the basis of the mathematical model which was established for the mechanism of operational equipment, the relations for determining the operation parameters of the hoisting and thrusting mechanisms (velocities of hoisting VH and thrusting V_T, forces of hoisting F_H and thrusting F_T) were obtained.

Table 1 shows the results of a computing experiment carried out in order to calculate the operation parameters for the main mechanisms of an EKG-20 excavator made by JSC «Uralmashplant».

The initial data for the calculation are:

- size of the excavator's work area (figure 3);
- paths of its bucket's motion (the top of the cutting edge – point *K*) as direct lines which follow the edges and the centre of the work area at the angle equal to the inclination of a face slope;
- velocity of excavation V_E = 1 m/s;
- rock resistance to excavation (tangent line) F_R^{τ} = 325 kN; gravity force of an empty bucket G_B = 400 kN; gravity force of a loaded bucket (bucket and rock) G_{B+R} = 700 kN and gravity force of the stick G_S = 300 kN.

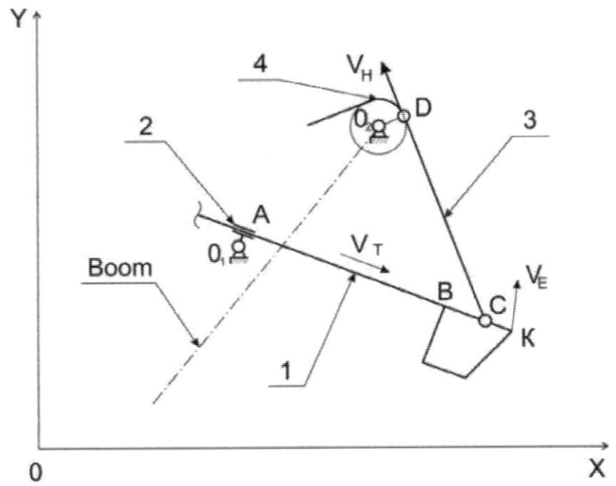

Figure 2. Schematics of the operational equipment mechanism: 1 – "stick-bucket" link; 2 – saddle bearing; 3 – segment of the hoisting rope; 4 – lead block; O_1, O_2 – stands; OX – datum level of the excavator; OY – rotation axis of its platform; V_H – velocity of hoisting; V_T – velocity of thrusting; V_E – velocity of excavation.

Table 1. The operation parameters for the main mechanisms of an EKG-20 excavator.

№	Initial data				Calculated data		
	X_K, m	Y_K, m	G_{B+R}, kN	V_H, m/s	V_T, m/s	F_H, kN	F_T, kN
				Initial path of motion (A)			
1	9	0	400	0.95	– 0.87	290	– 630
2	10.15	2	435	0.92	– 0.81	350	– 620
3	11.3	4	470	0.84	– 0.70	420	– 605
4	12.45	6	500	0.70	– 0.51	520	– 580
5	13.6	8	540	0.52	– 0.19	700	– 555
6	14.75	10	575	0.50	0.19	975	– 590
7	15.9	12	610	0.66	0.50	1270	– 700
8	17.05	14	650	0.80	0.70	1560	– 850
9	18.2	16	680	0.89	0.81	1810	– 1005
10	18.8	17	700	0.91	0.84	1910	– 1060
				Intermediate path of motion (B)			
11	13.5	0	400	0.88	– 0.64	610	– 310
12	14.65	2	435	0.83	– 0.51	640	– 300
13	15.8	4	470	0.76	– 0.34	680	– 280
14	16.95	6	500	0.70	– 0.12	710	– 230
15	18.1	8	540	0.66	0.11	780	– 155
16	19.25	10	575	0.65	0.32	830	– 40
17	20.4	12	610	0.62	0.50	885	140
18	21.55	14	650	0.50	0.63	930	420
19	22.7	16	680	0.22	0.72	1015	845
20	23.3	17	700	0.02	0.76	1125	1130
				Terminal path of motion (C)			
21	18	0	400	0.79	– 0.39	1020	110
22	19.15	2	450	0.73	– 0.25	1060	150
23	20.3	4	500	0.68	– 0.09	1100	210
24	21.45	6	550	0.61	0.07	1150	320
25	22.6	8	600	0.53	0.23	1210	480
26	23.75	10	650	0.42	0.38	1290	710
27	24.9	12	700	0.27	0.50	1420	1030

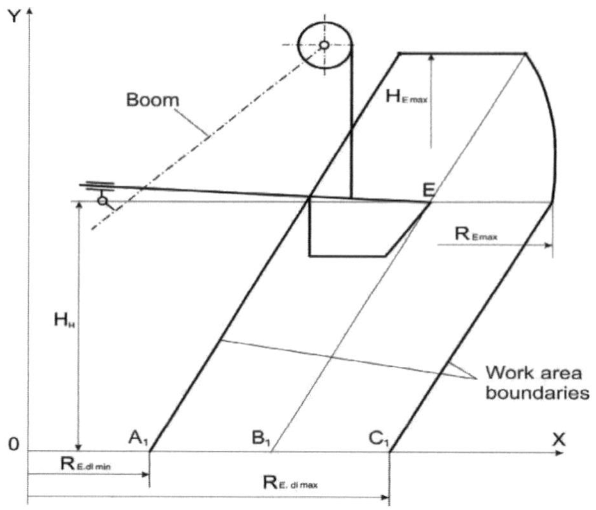

Figure 3. EKG-20 work area: A, B, C – paths of bucket motion; $H_{E.max}$, $H_{E.calc}$ – maximum and calculated heights of excavation; H_T – height of thrusting shaft axis; $R_{E.dl.min}$, $R_{E.dl.max}$ – minimum and maximum radiuses of excavation at the datum level of the excavator; $R_{E.max}$ – maximum radius of excavation.

The table demonstrates that the operating modes of the motors and the operation parameters of the main mechanisms change significantly within the excavator's work area depending on the position of its bucket. For example, when excavation is carried out in the bottom section of the work area at $Y_K \leq 6$, the motor of the hoisting mechanism is operating in traction mode, and the motor of the thrusting mechanism – both in traction mode with reverse rotation for the bucket following the paths A and B and in opposed action mode (braking – directions of F_T and V_T are opposite) for the bucket following the path C (figure 4).

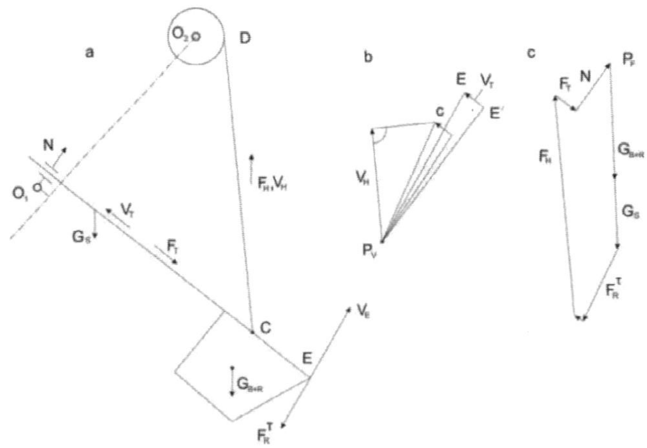

Figure 4. Diagram for determining operation parameters of the excavator with its bucket positioned in the bottom section of the work area (paths C): a – vector diagram of the mechanism, b – vector diagram of velocities, c – vector diagram of forces.

When excavation is carried out in the middle section of the work area at $6 \leq Y_K \leq 12$ m, the motor of the hoisting mechanism is operating in traction mode, and the motor of the thrusting mechanism – in three modes: traction mode with reverse rotation, dynamic braking mode for $F_T < 0$ (figure 5) and traction mode for the bucket following the path C (figure 6).

When excavation is carried out in the top section of the work area at $12 \leq Y_K \leq 17$ m, the motor of the hoisting mechanism is operating in traction mode, and the motor of the thrusting mechanism –

both in dynamic braking mode for the bucket following the path *A* and in traction mode for the bucket following the intermediate path *B*.

The operation parameters of the main mechanisms change in accordance with the operating modes of their motors and vary within a wide range [9] .

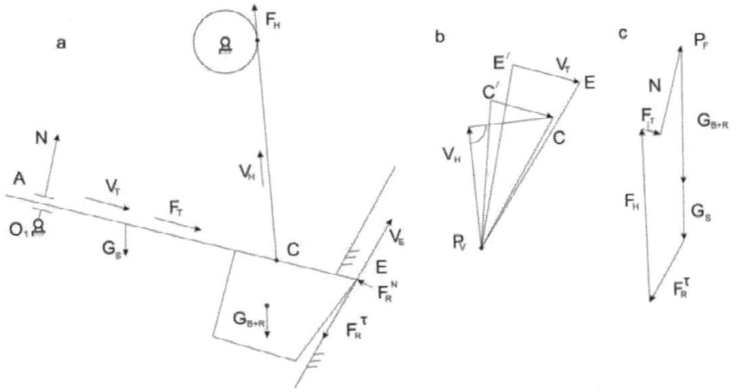

Figure 5. Diagram for determining operation parameters of the excavator with its bucket positioned in the middle section of the work area.

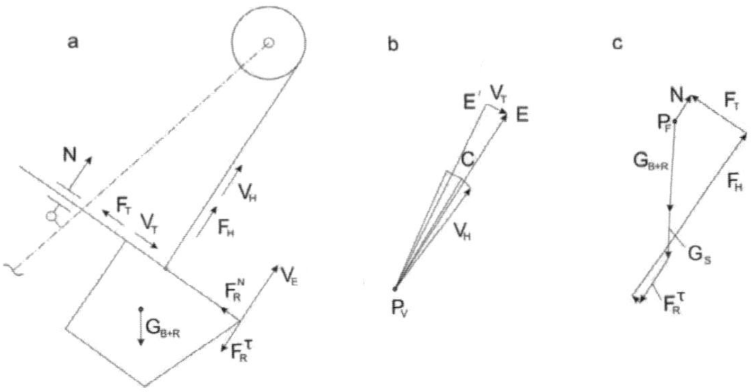

Figure 6. Diagram for determining operation parameters of the excavator with its bucket positioned in the middle section of the work area (path *A*).

The velocities of operating motions depend on the positions of the bucket in the work area (particularly, on the coordinate Y_K). With the exception of the bucket following its initial path (figure 7), the velocity of hoisting decreases with an increase in the height of excavation (Y_K), and the velocity of thrusting decreases for lower heights of excavation and increases with an increase of the height. The maximum values of those velocities are close to the velocity of excavation: $V_{H.max} = 0.95$ m/s и $V_{T.max} = 0.84$ m/s (for the initial path of motion).

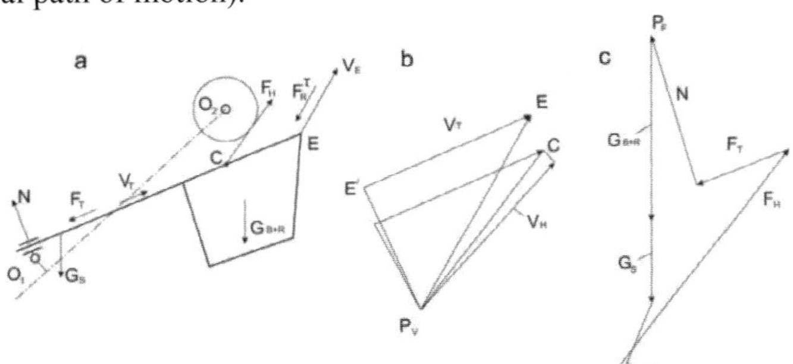

Figure 7. Diagram for determining operation parameters of the excavator with its bucket positioned in the top section of the work area (path A).

The forces of hoisting and thrusting increase with an increase in the height of excavation (with the exception of thrusting forces for the bucket following the intermediate path B). The maximum values of those forces are $F_{H.max}$ = 2.05 MN and $F_{T.max}$ = 1.26 MN. The maximum values referred in the specification for the excavator are considerably smaller. For instance, the value of the maximal hoisting force [10] – which determines the braking torque of the hoisting mechanism motor – $F_{H.max}$ = (1.25…1.43)$F_{H.calc}$ = 1350…1540 kN ($F_{H.calc}$ = 1080 kN is the force of hoisting with the height of excavation equal to the height of the thrusting shaft axis for the bucket following the intermediate path).

The increase in the forces of hoisting and thrusting with an increase in the height of excavation is caused by counteraction of the hoisting and thrusting motors (differently directed forces of hoisting and thrusting). In that case, the moment of forces which tilts the boom towards the excavator increases, which can lead to the boom swinging (jacking [11]) and, subsequently, returning (falling) into its initial position (which is accompanied by considerable dynamic loads).

4. Conclusion

In the process of excavation, the operating modes of the hoisting and thrusting motors and the operation parameters (velocity and forces) of the mechanisms change considerably depending on the position of the bucket in the excavator's work area.

In that case, controlling the process and maintaining the paths of motion which should be followed by the bucket and which correspond to the inclination of a face slope (rockfall) is quite a difficult task to handle by means of current tools (joysticks).

References

[1] Borshh-Komponiec L. V. Methodology of prompt assessment for open-pit hydraulic excavators // Mining industry. 1996. № **1**. pp. 29-37.
[2] Komissarov A. P., Lautenshlejger A. A., Suslov N. M. Assessment of energy parameters of hydraulic-excavator operational equipment // Heavy engineering industry. 1991. № **8**. pp. 25-29.
[3] Bules P. Operating efficiency of open-pit excavators with electromechanical and hydraulic drives for main mechanisms // Mining industry. 2014. № **6** (118). pp. 36-37.
[4] Visbek Z. et al. On efficiency of using open-pit hydraulic excavators // Mining industry. 1998. № **5**. pp. 25-29.
[5] Dudczak A. Excavators: theory and design. Warsaw: PWN, 2000.
[6] Geu Flores F., Kecskemethy A., Pottker A. Workspace analysis and maximal force calculation of a face-shovel excavator using kinematical transformers. 12th IFToMM World Congress, Besancon, June 18-21, 2007. 6 p.
[7] Frimpong S., Hu Y., Chang Z. Perfomance simulation of shovel excavators for earthmoving operations // In Summer in computer simulation conference (SCSC'03). 2003. pp. 133-38.
[8] Park B. Development of a virtual reality excavator simulator: a mathematical model of excavator digging and a calculation methodology. PhD Diss. Virginia Polytechnic Institute and State University. Blackburg, Virginia, USA, 2002. 223 p.
[9] Epifanov A. P., Malajchuk L. M., Gushhinskij A. G. Electric drive. Saint Petersburg: Lan, 2012. 400 p.
[10] Beljakov Yu. I. Planning excavation procedures. Moscow: Nedra, 1983.
[11] Podjerni R. Yu. Machinery of mines. Moscow: MGGU, 2007. 680 p.

Evaluating the performance of free-formed surface parts using an analytic network process

Xueming Qian, Yanqiao Ma and Dezhi Liang

School of Mechatronic Engineering, Xi'an Technological University, Xi'an, China

*Corresponding author's e-mail: bearqianxueming@163.com

Abstract. To successfully design parts with a free-formed surface, the critical issue of how to evaluate and select a favourable evaluation strategy before design is raised. The evaluation of free-formed surface parts is a multiple criteria decision-making (MCDM) problem that requires the consideration of a large number of interdependent factors. The analytic network process (ANP) is a relatively new MCDM method that can systematically deal with all kinds of dependences. In this paper, the factors, which come from the life-cycle and influence the design of free-formed surface parts, are proposed. After analysing the interdependence among these factors, a Hybrid ANP (HANP) structure for evaluating the part's curved surface is constructed. Then, a HANP evaluation of an impeller is presented to illustrate the application of the proposed method.

1. Introduction

As the global business competition environment becomes more intense, enterprises depend on new product development (NPD) for competitive advantage[1]. Most NPD budgets are committed during the design phase, before the actual work tasks take place. Adequate planning is one of the key factors required to satisfy project quality, to reduce financial and scheduled risks, and to help in the success of a project. The management of the planning phase of an NPD project is both significant and complicated. The evaluation of alternatives is the first basic step of product planning. The selection of alternatives can be used to direct the project planning process (e.g., project organization, budget, and schedule), which can determine the success of a project [2][3]. In the up-front design process, the systematic evaluation and verification of the design solution are critical since they could shift the product development paradigm from the traditional trial-and-error and heuristic know-how methods to more scientific calculation, analysis and simulation. Solution needs to be developed to ensure the efficient and accurate assessment and evaluation of design.

On the other hand, with the development of Group Technology and Mass Customization, the budget for common parts is reduced, while the budget rate for special parts rises sharply. These parts have a free-formed surface, which raises the difficulty of designing, manufacturing and testing, especially in the aviation industry, the mold industry, and the motor industry[4][5][6]. In this paper, multi-objects for free-formed surface part (FSP) evaluation are proposed, they are the designing performance, the manufacturing performance and the testing performance. Based on the multi-objects, a hybrid ANP model is established for FSP evaluation. The rest of the paper proceeds as follows. Section 2 provides a quick review of related literature. The ANP method is introduced in section 3. Then, the factors in FSP evaluation are analyzed in section 4. Established hybrid ANP evaluation structure is discussed in section 5. Section 6 evaluated impellers as a case study. Finally, section 7 concludes the paper.

2. Literature review

Many studies have evaluated product performance. P. Duverlie used an analogical method to evaluate design based on cost [7]. The analogical method allows for a cost-evaluation of a product compared with the cost of existing products. Hsu-Fang Hung evaluated design alternatives in collaborative development based on cost, time and quality. Product strategy is identified as the most important factor in this evaluation [8]. R. Sharma developed a system according to the criteria of time and cost, which allows the designer to estimate manufacturability and also explore different scenarios of parametric variation in the design[9]. Shehab and Abdalla proposed a manufacturing cost estimation tool for conceptual design using a CAD system in conjunction with a process planning system to estimate cost and generate initial plans[10]. The evaluation criteria mainly focused on individual development stages, such as: design, manufacture, quality, and testing. Today, evaluation of a individual development stage is deeply researched (e.g., manufacturing design, client-oriented design, and quality-oriented design).

The evaluation of a individual development stage mainly focuses on one object (for function, manufacturability or cost). In the design stage, the designers pour their energy into product function. Engineers think about how to realize the scenario in the manufacturing stage. In testing, personnel hope that the performance is easy to test and that the criteria are independent of each other. The criteria for these stages are interdependent, however. Subsequently, product design problems are decomposed into sub-problems, which evaluate the many aspects of a large problem. This step involves specialized knowledge from many fields that cannot be mastered by a single designer. Different aspects of a problem should be modeled and solved by different people who have appropriate core competencies [11][12]. Separate evaluation of the performances in a single development stage causes difficulty in evaluating general performance, especially when the product contains free-surface parts like the impeller, bodyworks, and mold pocket, since there are different objects to be matched in different stages.

Several methods have been proposed in the literature for multi-objective decision making problems. When considering intangible factors, some authors suggest using an analytic hierarchy process (AHP). The analytic hierarchy process (AHP) was developed by Thomas Saaty in the early 1970s[13]. The strength of the AHP approach lies in its ability to hierarchically structure a complex, multiattribute, multiperson and multiperiod problem. In addition, it can also handle both qualitative (through representing qualitative attributes in terms of quantitative values) and quantitative attributes. Pairwise comparisons of the elements (usually, alternatives and attributes) can be established using a scale that indicates the strength with which one element dominates another with respect to a higher level element. This scaling process can be translated into priority weights (scores) for a comparison of alternatives. However, AHP is based on an independence assumption, ignoring the interdependence fact in solving those problems.

Hence, Satty modified the AHP into an ANP approach to counterweigh the demerit[14]. The ANP captures the interdependence of decision criteria and improves the limitation of AHP. There are many authors who have recently applied this technique in MCDM problems. For example, Meade et al. presents an overall systemic methodology, which includes ANP and utility theory, for making a business case for strategic alliance formation and justifying the conclusion that strategic relationships are a critical step in the case firm[15]. Sarkis integrated the environmentally conscious business practice elements and attributes into a strategic assessment and applied the systems using ANP[16]. This integration highlighted the importance of the natural environment in everyday and long term organizational practices and processes at a level that is unparalleled since the start of the industrial solution. Meade and Sarkis showed what lies in the linkage of disparate strategic logistics and system issues in a single systematic study framework[17]. Here, ANP not only provides solutions to strategic logistics alternative selection, but it also provides a structure for an organization to develop and enhance logistics strategy. Wu and Lee used ANP to ensure the successful implementation of knowledge management[18]. They also raised the critical issue of how companies can better evaluate and select a favorable knowledge management strategy before that implementation. Thus, the ANP research method is most suitable and valid for this MCDM problem and interdependency relationships.

In this paper, we deal with FSP evaluation as a multi-objective decision making problem. First, the paper proposes three phases to build network systems for comprehensive evaluation. The authors' analyzed the evaluating factors of free-formed surface parts (FSP). Second, all the factors are classified into three layers. Third, the second layer is treated as the network layer, and a hybrid ANP model is established for FSP evaluation.

3. Principles of the ANP Method

3.1. The typical structure of an analytic network process

The typical network architecture is made up of two parts (shown as Figure 1). These parts include: (1) a control layer and (2) a network layer. In the control layer there are some decision-making criteria, like $P_1 \cdots, P_i \cdots P_m$. These elements are only controlled by the goal element, and can be divided into sub-criterion according to need. This means that the control layer can have many goal elements or only one. The other part is the network-layer. It contains some element clusters, such as $\{C_1 \cdots, C_i \cdots C_n\}$, here, C_i denotes element cluster $\{e_{i1}, e_{i2}, \cdots, e_{in}\}$. The weights of the elements in the control-layer can be obtained by the AHP. In the network layer, they must be obtained via super-matrix.

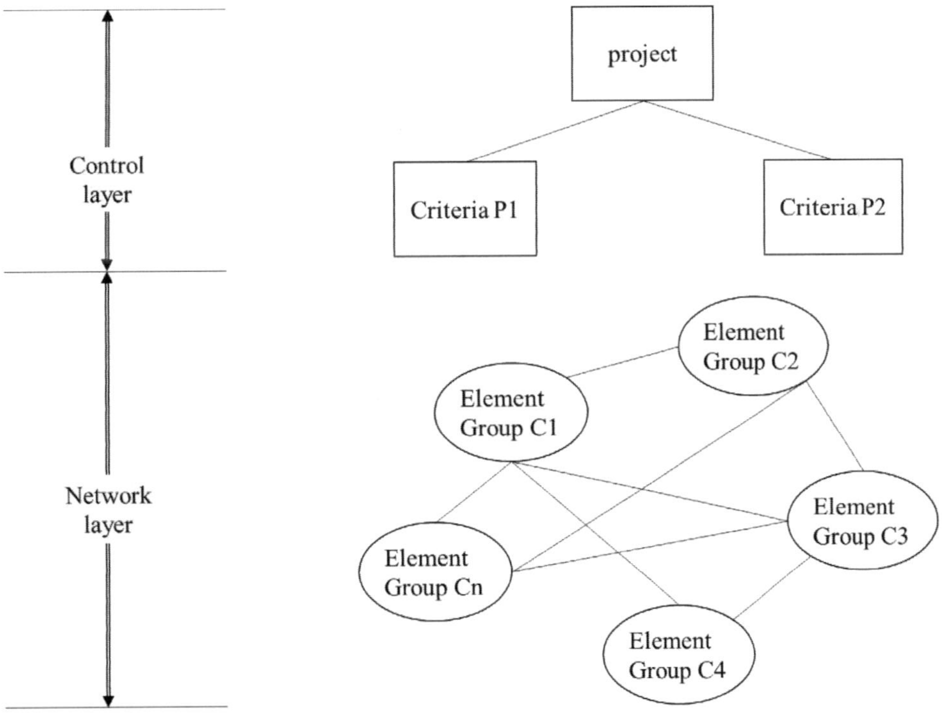

Figure 1. ANP structure.

The Analytic Network Process (ANP) provides a general framework for dealing with decisions without making assumptions about the independence of higher-level elements from lower level elements or the independence of the elements within a level (as in a hierarchy). In fact, the ANP uses a network without a need for special levels.

IOP Conf. Series: Materials Science and Engineering **323** (2018) 012017 doi:10.1088/1757-899X/323/1/012017

4. Analyse the Factors in FSP Evaluation

4.1. Evaluation criteria

These evaluation criteria are management focused and mainly evaluate the cost, delivery and quality, while it is difficult to control the result without controlling the process. In order to better evaluate the FSP, the criteria from the three development stages are proposed here.

Evaluation criteria come from three phases: (1) design process, (2) manufacturing process, and (3) test process.

The evaluation factors for the design performance of FSP include: S_1 hydromechanics, S_2 aesthetics, S_3 kinetics, S_4 toughness, S_5 other performance, S_6 fairing S_7 , free-formed surface shape, S_8 smoothing, S_9 the fairing of key section curve, S_{10} Gauss Curvature, S_{11} principal Curvature, S_{12} inflexion and Variable torsion Point, S_{13} mean curvature, S_{14} parameter continuity, S_{15} Geometric continuity, S_{16} curvature, S_{17} inflexion, S_{18} Variable torsion Point, S_{19} second derivative point, S_{20} control point, S_{21} weight factor, S_{22} node vector, S_{23} data points, S_{24} the number of data point, S_{25} the distribution of data point.

The reachable matrix among the criteria is established using a normative method after the matrix is hierarchically ordered, a sub-triangular matrix is elicited as shown in Figure 2.

	S_1	S_2	S_3	S_4	S_5	S_6	S_7	S_8	S_9	S_{10}	S_{11}	S_{12}	S_{13}	S_{14}	S_{15}	S_{16}	S_{17}	S_{18}	S_{19}	S_{20}	S_{21}	S_{22}	S_{23}	S_{24}	S_{25}
S_1	1	0	0	0	0	0	0	0	0	0	0	0	0	0	0	0	0	0	0	0	0	0	0	0	0
S_2	0	1	0	0	0	0	0	0	0	0	0	0	0	0	0	0	0	0	0	0	0	0	0	0	0
S_3	0	0	1	0	0	0	0	0	0	0	0	0	0	0	0	0	0	0	0	0	0	0	0	0	0
S_4	0	0	0	1	0	0	0	0	0	0	0	0	0	0	0	0	0	0	0	0	0	0	0	0	0
S_5	0	0	0	0	1	0	0	0	0	0	0	0	0	0	0	0	0	0	0	0	0	0	0	0	0
S_6	0	0	0	0	0	1	0	0	0	0	0	0	0	0	0	0	0	0	0	0	0	0	0	0	0
S_7	1	1	1	0	0	0	1	0	0	0	0	0	0	0	0	0	0	0	0	0	0	0	0	0	0
S_8	1	1	1	0	0	0	0	1	0	0	0	0	0	0	0	0	0	0	0	0	0	0	0	0	0
S_9	0	0	0	0	0	1	0	0	1	0	0	0	0	0	0	0	0	0	0	0	0	0	0	0	0
S_{10}	0	0	0	0	0	1	0	0	0	1	0	0	0	0	0	0	0	0	0	0	0	0	0	0	0
S_{11}	0	0	0	0	0	1	0	0	0	0	1	0	0	0	0	0	0	0	0	0	0	0	0	0	0
S_{12}	0	0	0	0	0	1	0	0	0	0	0	1	0	0	0	0	0	0	0	0	0	0	0	0	0
S_{13}	0	0	0	0	0	1	0	0	0	0	0	0	1	0	0	0	0	0	0	0	0	0	0	0	0
S_{14}	0	0	0	0	0	1	0	0	0	0	0	0	0	1	0	0	0	0	0	0	0	0	0	0	0
S_{15}	0	0	0	0	0	1	0	0	0	0	0	0	0	0	1	0	0	0	0	0	0	0	0	0	0
S_{16}	0	0	0	0	0	1	0	0	1	0	0	0	0	0	0	1	0	0	0	0	0	0	0	0	0
S_{17}	0	0	0	0	0	1	0	0	1	0	0	0	0	0	0	0	1	0	0	0	0	0	0	0	0
S_{18}	0	0	0	0	0	1	0	0	1	0	0	0	0	0	0	0	0	1	0	0	0	0	0	0	0
S_{19}	0	0	0	0	0	1	0	0	1	0	0	0	0	0	0	0	0	0	1	0	0	0	0	0	0
S_{20}	0	0	0	0	0	1	0	0	1	0	0	0	0	0	0	1	0	0	0	1	0	0	0	0	0
S_{21}	0	0	0	0	0	1	0	0	1	0	0	0	0	0	0	1	0	0	0	0	1	0	0	0	0
S_{22}	0	0	0	0	0	1	0	0	1	0	0	0	0	0	0	1	0	0	0	0	0	1	0	0	0
S_{23}	0	0	0	0	0	1	0	0	1	0	0	0	0	0	0	1	0	0	0	0	0	0	1	0	0
S_{24}	0	0	0	0	0	1	0	0	1	0	0	0	0	0	0	1	0	0	0	0	0	0	1	1	0
S_{25}	0	0	0	0	0	1	0	0	1	0	0	0	0	0	0	1	0	0	0	0	0	0	1	0	1

Figure 2. Criteria in the design process.

According to the above matrix (Figure 2), S_1, S_2, S_3, S_4, S_5 and S_6 belong to the first layer. S_7, S_8, S_9, S_{10}, S_{13}, S_{14} and S_{14} belong to the second layer. S_{16}, S_{17}, S_{18} and S_{19} belong to the third layer, along with S_{20}, S_{21}, S_{22} and S_{23}. S_{24} and S_{25} belong to the sixth layer.

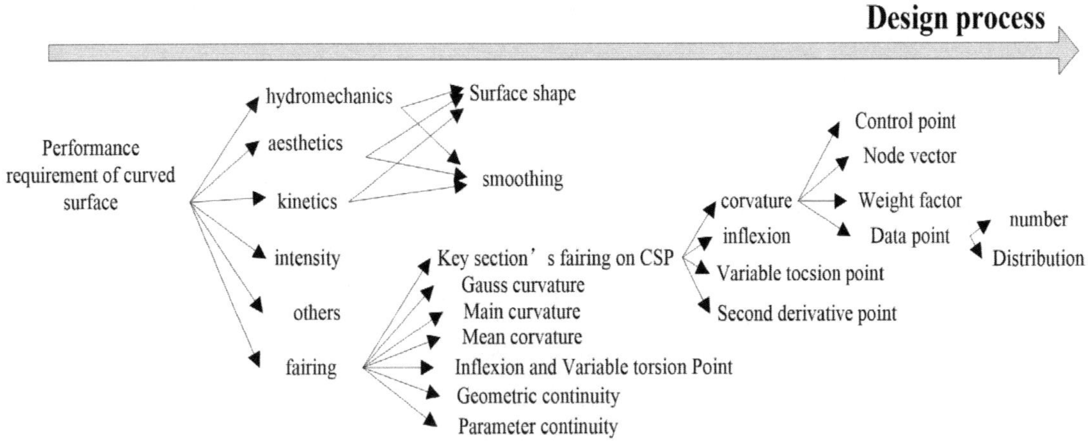

Figure 3. Criteria in the design process.

First, the above numerals elements were arrayed according to the mentioned hierarchy. The directed graph (DG) according to element "1" in the interchange place was then drawn. The hierarchical DG was then obtained, which presents the direct relationship of elements in different layers, similar to the above chart (Figure 3). The chart is a gradually specific design process from left to right. The criteria model in manufacturing and testing are set up at the same way, which is displayed in the following picture Figure 4.

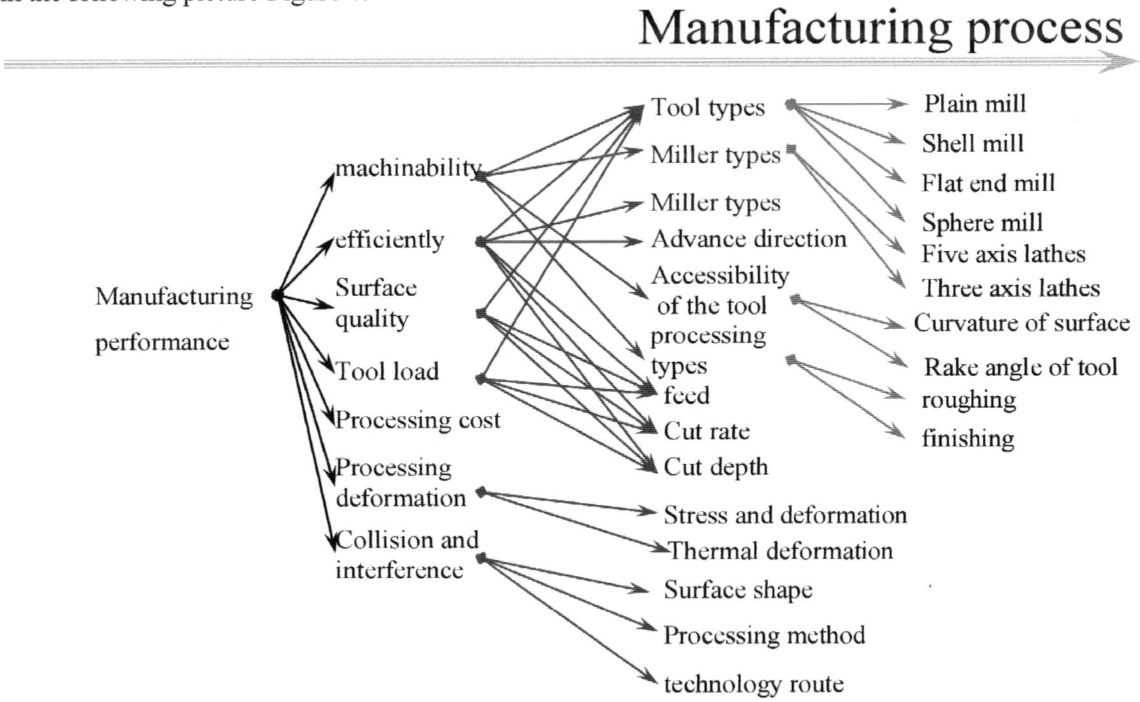

Figure 4. Criteria for manufacturing process.

The multi-level structured graph is shown in the following paragraph Figure 5.

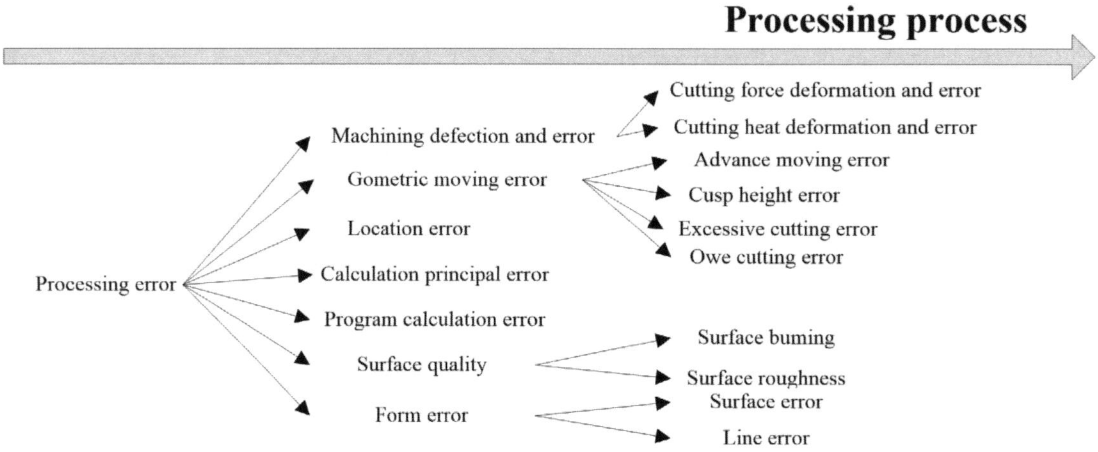

Figure 5. Criteria for processing error.

5. Hybrid ANP Structure for FSP Evaluation

This hierarchical system simply and intuitively describes the relationship of various indicators (factors in the lower classes dominated by the upper classes). It does not reflect interactions and inter-dependencies in the same class. ANP is an effective method to show the interdependence among criteria [19].

The ANP provides a more generalized model in decision-making without making assumptions about the independency of the elements within their own level. However, to evaluate the FSP, there are sixty-two criterions. Thus, ANP reduces the efficiency of the evaluation. Therefore, it is necessary to adjust the hierarchical system of the FSP and establish a mixed-network model according to the typical structure of ANP.

In this paper, according to the factors discussed in section 3, we treated the second layer criteria as the network layer, while the third layer and fourth layer were still in a hierarchical structure. We then set up a hybrid ANP structure.

5.1. Criteria and criteria clusters in the network

The control layer includes three criterion clusters: P_1, P_2, and P_3. P_1 is the design performance of the free-formed surface. P_2 is the manufacturing performance. P_3 is the testing performance. The criteria in the paper are mainly the elements dominated by the control layer in the three processes of free-formed surface design, process, and test, which are classified into three groups according to the elements. The three element groups are: C_1, C_2, and C_3. C_1 includes e_{11} hydromechanics, e_{12} aesthetics, e_{13} kinetics, e_{14} intensity, e_{15} other performance, and e_{16} fairing. C_2 includes, e_{21} machinability, e_{22} processing efficiency, e_{23} surface quality, e_{24} tool load, e_{25} processing cost, e_{26} processing deformation, and e_{27} collision and interference. C_3 includes e_{31} machining deflection and error, e_{32} geometric moving error, e_{33} programs calculation error, e_{34} calculation principal error, e_{35} surface quality, e_{36} location error, and e_{37} form error.

5.1.1. The hybrid network modelling evaluation system on the basis of hybrid -ANP.

As shown in Figure 6, the satisfaction of the CSP is the overall object. The three criteria are: design performance, manufacturing performance, and testing performance. The control layer is formed by these three criteria.

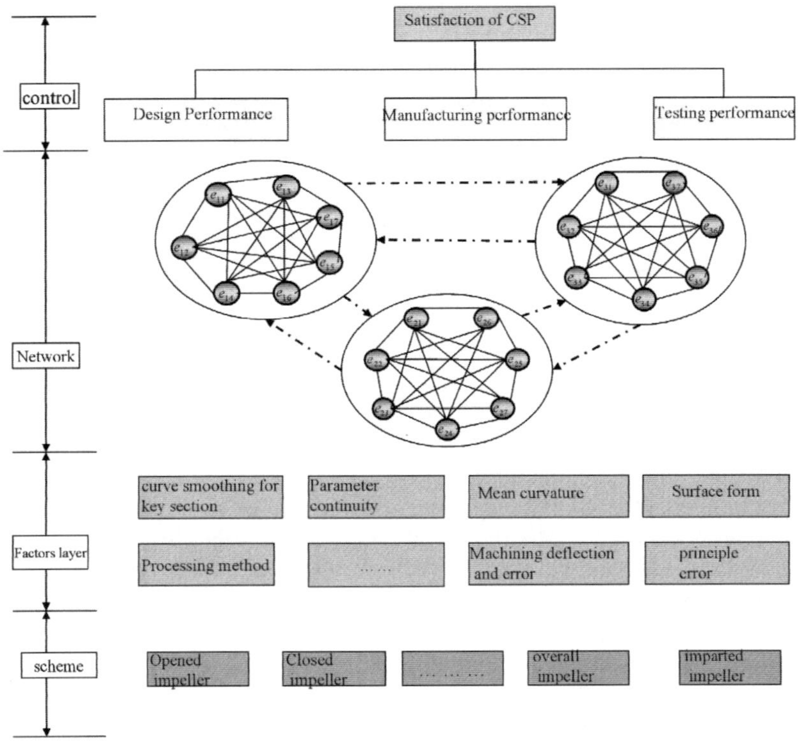

Figure 6. Hybrid ANP structure.

The three criterion clusters are formed by the elements of the second layer (design, process, and test). In the third and fourth layer, design, manufacturing, and test are named as factor levels. The framework of the factor level is still the hierarchical structure. The schemes are treated as the fourth layer, where mixed evaluation architecture is established.

5.2. Determining the total sort eigenvector

Table 1 is the sort eigenvector that is calculated according to the limit relative sort eigenvector in the criteria of the control layer and the criteria of the Network layer. In order to compare the evaluation results, the sort eigenvector is calculated by AHP in the table at the same time.

Table 1. Total sort eigenvector.

Sort eigenvector		Sort eigenvector on the criteria						Overall sort eigenvector	
criteria / Elements in the network layer		P_1 W_1=0.4819		P_2 W_2=0.4462		P_3 W_3=0.0719		Project	
		ANP	AHP	ANP	AHP	ANP	AHP	ANP	AHP
C_1	e_{11}	0.0574	0.1146	0.0574		0.0542		0.0572	0.0552
	e_{12}	0.0577	0.1126	0.0537		0.0785		0.0574	0.0543
	e_{13}	0.0572	0.1795	0.0546		0.0761		0.0574	0.0865
	e_{14}	0.0627	0.2027	0.0637		0.0707		0.0637	0.0977
	e_{15}	0.0575	0.0478	0.0375		0.0425		0.0475	0.0230
	e_{16}	0.0481	0.3428	0.0481		0.0481		0.0481	0.1652

C_2	e_{21}	0.0571	0.0571	0.2621	0.0571		0.0571	0.1169
	e_{22}	0.0509	0.0479	0.0899	0.0695		0.0509	0.0401
	e_{23}	0.0572	0.0602	0.2654	0.0386		0.0572	0.1184
	e_{24}	0.0412	0.0442	0.1160	0.0226		0.0412	0.0518
	e_{25}	0.0579	0.0429	0.0561	0.0119		0.0479	0.0250
	e_{26}	0.0445	0.0495	0.1104	0.0134		0.0445	0.0493
	e_{27}	0.0345	0.0305	0.1001	0.0593		0.0345	0.0447
C_3	e_{31}	0.0391	0.0391		0.0391	0.1156	0.0391	0.0083
	e_{32}	0.0473	0.0753		0.1507	0.1395	0.0673	0.0100
	e_{33}	0.0453	0.0403		0.0068	0.0539	0.0403	0.0039
	e_{34}	0.0472	0.0602		0.0361	0.0779	0.0522	0.0056

6. Real World Case Study

The impeller is a complex product with free-formed surface parts. Figure 7 shows two impeller schemes for one company. According to the ability in design, manufacturing and testing of the company, the performance of the opened impeller and closed impeller are comprehensively evaluated in accordance with the comprehensive sort eigenvector in table 1. The overall scores are calculated by ANP and AHP. If the dependency is omitted, the ANP model becomes an AHP model. The opened impeller scheme scores 5.1971441 by the ANP method and 5.521217 by the AHP method. The closed impeller scheme scores 5.388804 by the ANP method and 5.53628 by the AHP method. The ANP method is more distinct than the AHP. Furthermore, the ANP method takes the dependence of the element in the same layer into account. It is also reasonable at distributing the weight among criteria. The scores level is lower than the analytical method in evaluating the same scheme.

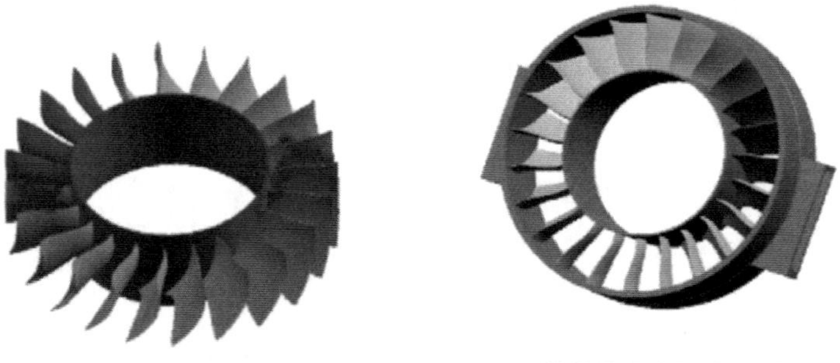

(a) Closed loop impeller (b) Open loop impeller

Figure 7. Two kinds of impeller.

7. Conclusion

This paper presented various evaluation criteria for the design, manufacturing and testing of an FSP. According to the dominance relations and interdependence among the criteria, we established a hybrid Network evaluation system based on HANP and gave a comprehensive evaluation for the impeller scheme. Compared with the traditional method AHP, HANP considered the interdependence among the criteria in same layer, making the evaluation results more reasonable and the scheme easier to distinguish. In the evaluation system, the scoring system still uses a decimalist, which is constitutionally based on the score of specialists from different areas. This scoring system is difficult and inefficient at avoiding of subjectivism of specialists. Thus, the scoring system for design, manufacturing, and testing is the focus of further study.

IOP Conf. Series: Materials Science and Engineering **323** (2018) 012017 doi:10.1088/1757-899X/323/1/012017

References

[1] Hicks BJ, Cully SJ, Allen RD, and Mullineux G 2002 A framework for the requirements of capturing, storing and reusing information and knowledge in engineering design. *Int J Inf Manage* 22:263–80

[2] Lee E,Woo S, Sasada T 2001 The evaluation system for design alternatives in collaborative design. *Automat Construction* 10:295–301

[3] Dvir D, Raz T, and Shenhar AJ 2003 An empirical analysis of the relationship between project planning and project success. *Int J Project Manage* 21:89–95

[4] Chen YM, Liu JJ 1999 Cost effective design for injection molding. *Robot Comput Integr Manuf* 15:1–21

[5] W. El-Mehalawi and R. A. Miller 1998 A case-based approach for part cost estimation in net shape manufacturing. *Transactions of NAMRIC/SAE* 26: 37–42

[6] H. Wang, X. H. Zhou and X.-Y. Ruan 2003 Research on Injection Mould Intelligent Cost Estimation System and Key Technologies. *Int J Adv Manuf Technol* 3:215–22

[7] P. Duverlie 1997 Cost estimating during design step: parametric method versus Case based reasoning method. *Universite' de Valenciennes*

[8] Hsu-Fang Hung, Hsing-Pei Kao, and Kuo-Cheng Ku 2007 Evaluation of design alternatives in collaborative development and production of modular products. *Int J Adv Manuf Technol* 33: 1065–076

[9] R. Sharma. J. X. Gao 2007 A knowledge-based manufacturing and cost evaluation system for product design/re-design. *Int J Adv Manuf Technol* 33: 856–65

[10] Shehab EM, Abdalla HS 2001 Manufacturing cost modeling for concurrent product development. *Robot Comput-Integr Manuf* 17 (4): 341–53

[11] Eppinger, S. D., Whitney, D. E., Smith, R. P. and Gebala, D. A 1994 A model-based method for organizing tasks in product development. *Research in Engineering Design* 6: 1–13

[12] Krishnamachari, R. S. and Papalambros, P. Y. 1996. Optimal hierarchical decomposition synthesis using integer programming. *Proceedings of 1996 ASME Design Engineering Technical Conferences and Computersin Engineering Conference*. Irvine, CA

[13] Saaty, T.L 1988 The Analytic Hierarchy Process. *University of Pittsburgh*

[14] Saaty T.L 1996 Decision making with dependence and feedback: the analytic network process. RWS, Pittsburgh

[15] Meade LM, Liles DH, and Sarkis J 1997 Justifying strategic alliances and partnering: a prerequisite for virtual enterprising. Omega 25 (1):29–42

[16] Sarkis J 1998 Evaluating environmentally conscious business practices. *Eur J Oper Res* 107:159–74

[17] Meade LM, Sarkis J 1998 Strategic analysis of logistics and supply chain management systems using analytical network process. *Logist Transp Rev* 34 (3):201–15

[18] Wen Wu, Yu-Ting Lee 2007 Selecting knowledge management strategies by using the analytic network process. *Expert Systems with Applications* 32 (3):841-847

[19] Saaty T.L 2001 Decision Making with Dependence and Feedback—The Analytic Network Process. 2nd ed., Pittsburgh, USA: RWS Publications

IOP Conf. Series: Materials Science and Engineering **323** (2018) 012018 doi:10.1088/1757-899X/323/1/012018

Synthesis, morphological, electromechanical characterization of $(CaMgFe_x)Fe_{1-x}Ti_3O_{12-\delta}$/PDMS nanocomposite thin films for energy storage application

Ashis Tripathy[1], Priyaranjan Sharma[2] and Narayan Sahoo[3]
[1]Department of Electronics and Communication Engineering, JK Lakshmipat University, Jaipur, India.
[2]Department of Mechanical Engineering, JK Lakshmipat University, Jaipur, India.
[3]Department of Electrical Engineering, JK Lakshmipat University, Jaipur, India.

*E-mail: ashisbidyarthi@gmail.com

Abstract. At the present time, flexible and stretchable electronics has intended to use the new cutting-edge technologies for advanced electronic application. Currently, Polymers are being employed for such applications but they are not effective due to their low dielectric constant. To enhance the dielectric properties of polymer for energy storage application, it is necessary to add ceramic material of high dielectric constant to synthesize a polymer-ceramic composite. Therefore, a novel attempt has been made to enhance the dielectric properties of the Polydimethylsiloxane (PDMS) polymer by adding $(CaMgFe_x)Fe_{1-x}Ti_3O_{12-\delta}$ ($0<x<1$, $0<\delta<6$) (CMFTO) nanocomposite ceramic powder. The newly developed CMFTO2/PDMS composite based thin film shows a higher dielectric constant (ε') value (~350), extremely low tangent loss ($\tan\delta$) (< 0.002) and excellent flexibility (static modulus ≈ 0.27 MPa and elongation > 90%), which can make it a potential material for advanced flexible electronic devices, energy storage and biomedical applications.

1. Introduction

Nowadays, polymer ceramic composites are gaining more and more interest of researchers due to versatility in the field of electronic materials, particularly in acoustic emission sensors, electronic packaging materials, energy-storage super capacitors and integrated decoupling capacitors[1,2]. Electroceramics are brittle in nature, exhibits medium dielectric strength and require high processing temperature; however, it exhibits high dielectric constant. Whereas polymers are flexible in nature followed by high dielectric breakdown voltage and low processing temperature, however they exhibit low dielectric constant. To obtain the collective benefits of polymer and electroceramic, it is necessary to develop polymer ceramic composite. Polymer ceramic composites can act as good dielectric materials. Therefore, it is necessary to develop suitable polymer ceramic composite in order to achieve better dielectric properties like dielectric constant, loss tangent and impedance at various frequency range by simply changing composite composition. These composites have good mechanical properties coupled with easy processing and can be formed into any typical shape through a simple molding process. Moreover, a suitable polymer matrix can increase both rigidity and flexibility of ceramic composites for numerous advanced electronic applications [3]. Previous researchers [4,5] have attempted to develop various polymer ceramic composites to enhance the dielectric properties of

IOP Conf. Series: Materials Science and Engineering **323** (2018) 012018 doi:10.1088/1757-899X/323/1/012018

existing polymer. Still, there is a scope to enhance the dielectric constant of polymer by adding nanoceramic composite for several energy storage applications.

In this current study, synthesis of $(CaMgFe_x)Fe_{1-x}Ti_3O_{12-\delta}$/PDMS based composite flexible films and modulation of their electrical and dielectric properties are reported. And, various performance attributes of polymer ceramic composite such as surface microstructure, electrical resistivity, mechanical flexibility and dielectric characteristics were analyzed.

2. Experimental section

2.1. Synthesis of $(CaMgFe_x)Fe_{1-x}Ti_3O_{12-\delta}$ nanocomposite and flexible film

The empirical formulas of the nanocomposite used for the current study were $CaMgFe_{0.33}Ti_3O_{12}$ (CMFTO1) and $CaMgFe_{0.67}Ti_3O_{12}$ (CMFTO2). The initial raw materials used for the study were Calcium Oxide (CaO), Magnesium Carbonate ($MgCO_3$), Hematite (Fe_2O_3), and Titania (TiO_2). Initially, these oxide powders were mixed with 70% ethyl alcohol in a ball milling machine using Zirconia balls. Then, mixed slurry was placed in an oven at 150 °C to remove the moisture and alcohol. The dry powder was sintered at 1050 °C for 16 h to form desired nanocomposite. Then, the sintered powder of CMFTO1 and CMFTO2 were ball milled into fine powder form and mixed with PDMS. To enhance crosslinking in PDMS chain, a curing agent was added to the mixture of CMFTO and PDMS. At 3000 rpm, CMFTO/PDMS mixture was spin coated on a glass container for 20 s under vacuum environment to form a thin film. To remove the micro-bubbles from the prepared thin films, they were kept in a self-drying vacuum pump under the pressure of 500 mbar for a period of 10 min. At last, flexible film of CMFTO/PDMS composite were obtained by curing at a temperature of 60 °C for 5 h. The synthesis process of the CMFTO nanocomposite and its PDMS based flexible composite film has been shown in figure 1.

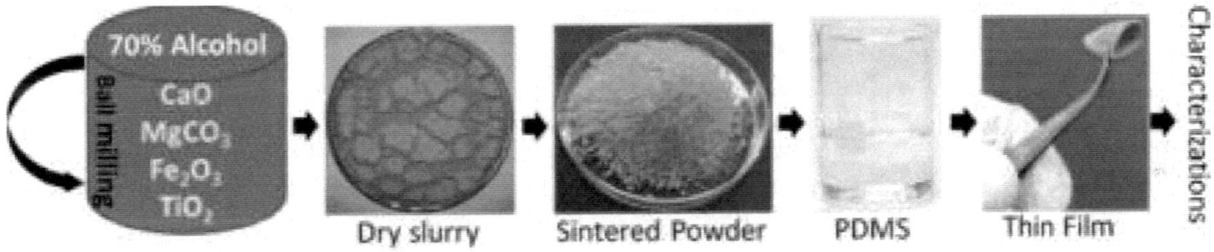

Figure 1. Synthesis process of CMFTO nanocomposites and CMFTO/PDMS thin films.

2.2. Characterizations

The impedance and dielectric characteristics of composite films were measured with the help of an LCR meter (3532-50 LCR Hi tester, Hioki, Ueda, Japan). The dispersion of filler in the polymer matrix and surface morphology of polymer ceramic composite were studied by using field emission scanning electron microscope (FESEM). A universal testing machine (5848, InstronMicro Tester) was used to measure the flexibility of the PDMS and CMFTO/PDMS composite films.

3. Results and discussion

3.1. Microstructural studies

FESEM images of the CMFTO2/PDMS and CMFTO1/PDMS composite films were shown in figure 2(a) and 2(b). The flexible CMFTO1/PDMS and CMFTO2/PDMS composite films show good particle matrix bonding at the polymer-ceramic interface. Inset of figures 2(a) and 2(b) shows the digital image of CMFTO2/PDMS and CMFTO1/PDMS composite film respectively. Uniform dispersion of ceramic composite particles in PDMS has been observed in the both the FESEM images. Due to the reunion mechanism, CMFTO nanocomposite agglomerate in certain areas which has been highlighted by the yellow circle in micrograph as shown in figure 2(a-b) .

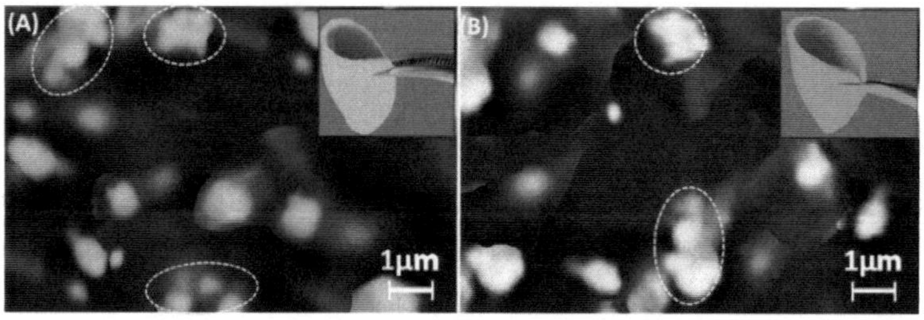

Figure 2. Field emission scanning electron microscope (FESEM) images of (a) CMFTO2/PDMS and (b) CMFTO1/PDMS composite films.

3.2. Impedance and dielectric spectroscopy

The frequency dependent real (Z') and imaginary (Z") parts of the impedance at room temperature for both the CMFTO/PDMS samples are depicted in figure 3(a) and 3(b), respectively. From the figure 3(a), it has been observed that at the low frequency range, Z' shows a larger value and it decreases significantly with increased frequency. It may be due to polarization effect in the samples. This decreasing trend in Z'-value shows a prospective of enhancement in the alternating current (AC) conductivity (σ_{ac}) with an increase in frequency. It has been also found that the impedance value decreases due to increase of Fe concentration. In higher frequency range, the Z' merges together, which gives a hint of probable discharge of space charge and a subsequent dropping of the barrier characteristics in the polymer ceramic composite [6].

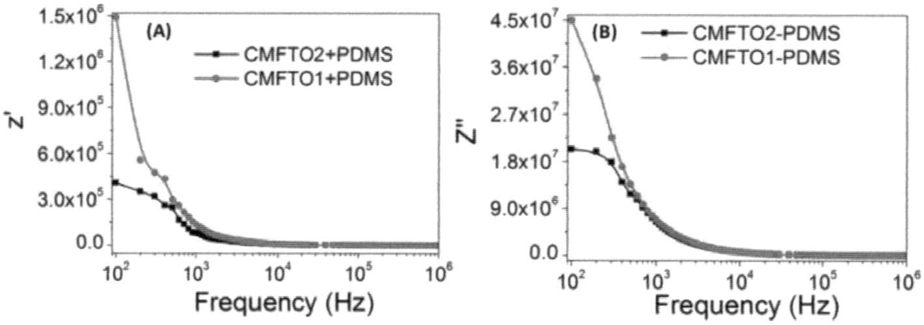

Figure 3. Variations of (a) real (Z') impedance (b) imaginary (Z") impedance, of the CMFTO1/PDMS and CMFTO2/PDMS composite films as a function of frequency in logarithmic scale at 25°C.

The appearance of peak in the Z" with frequency (see figure 3b) recommends the presence of relaxation phenomena of the material. Due to increasing of Fe concentration, more broadening of peak is observed. The unequal expansion of the peak recommends the existence of electrical processes in the composite film with distribution of relaxation time [6]. This broadening of peak after addition of Fe may be due to some relaxation phenomena caused by hopping of electrons/oxygen ion vacancy/defects among the available localized sites [6]. Gathering of space charge of the ceramic-polymer nanocomposite may be the possible reason for the monotonous decrease of Z"-value with increasing frequency after maxima.

The changes in dielectric constant (ε') and dielectric loss (ε'') for CMFTO1/PDMS and CMFTO2/PDMS composite films are shown in figure 4(a) and 4(b) as a function of frequency in logarithm scale at 25 °C. The value of ε' was increased with the increased concentration of Fe but decreased with increase in frequency as shown in figure 4(a). The concentration variation and frequency dependency of dielectric constant are more pronounced in the lower frequency range ($f<10^3$ Hz). Enhancement of interfacial polarization is the main cause for the significant improvement of ε' at the lower frequency range. In the same fashion, the ε" was also increased with the increased concentration of Fe (figure 2b). The peaks in dielectric loss vs. frequency plot indicate the occurrence of some relaxation process in the system. Due to the extremely high ε' value of CMFTO2/PDMS composite, it could be used for various capacitive and electric energy storage applications.

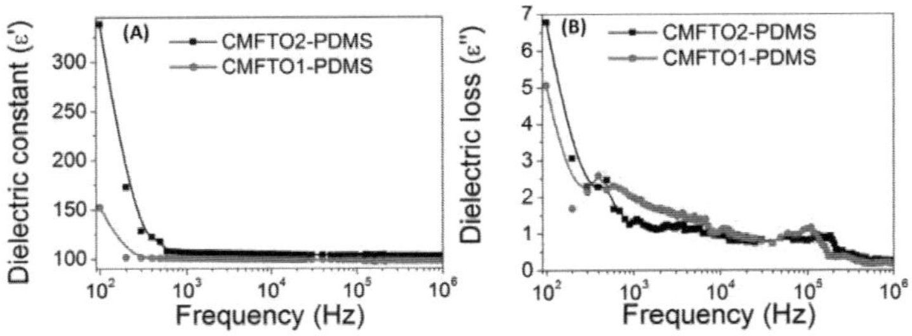

Figure 4. The variations of (a) Dielectric constant (ε′), (b) Dielectric loss (ε″) versus frequency plots of the CMFTO1/PDMS and CMFTO2/PDMS composite films.

Frequency dependent dielectric constant (ε ′) and tan δ response for CMFTO2/PDMS and CMFTO1/PDMS composite films are shown in figure 5(a-b) respectively. CMFTO2/PDMS film shows a higher dielectric value (~350) and extremely low tanδ (< 0.002). Extremely high ε′ and low tanδ value of CMFTO2/PDMS composite film can make it a highly prospective candidate for high energy storage capacitor applications. In the figure it is also observed that the CMFTO1/PDMS composite film shows lower ε′ (i.e., ~160) than the CMFTO2/PDMS composite film, but effectively greater than the pure PDMS (ε′ ~10)) and PDMS based nanocomposites [7].

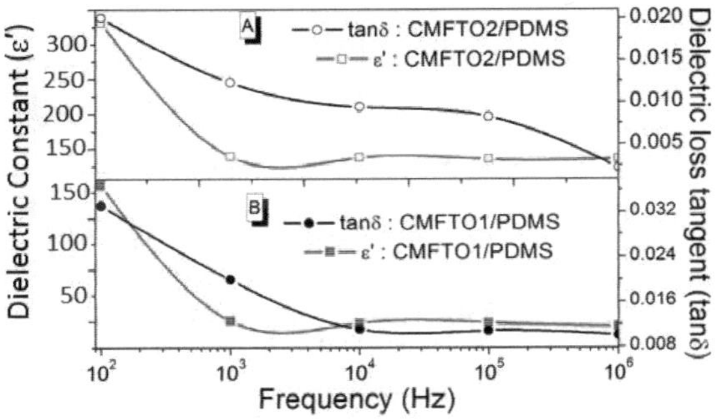

Figure 5. A typical comparison in ε′ and tanδ of (a) CMFTO2/PDMS and (b) CMFTO1/PDMS composite film at different frequency.

AC conductivity (σ_{ac}) vs. frequency response of CMFTO1/PDMS and CMFTO2/PDMS composite film has been shown in the figure 6. From the response, it was observed that σ_{ac} of both the ceramic-polymer composite slowly increased up to log(*f*) = 4.4 Hz and increased faster after this frequency.

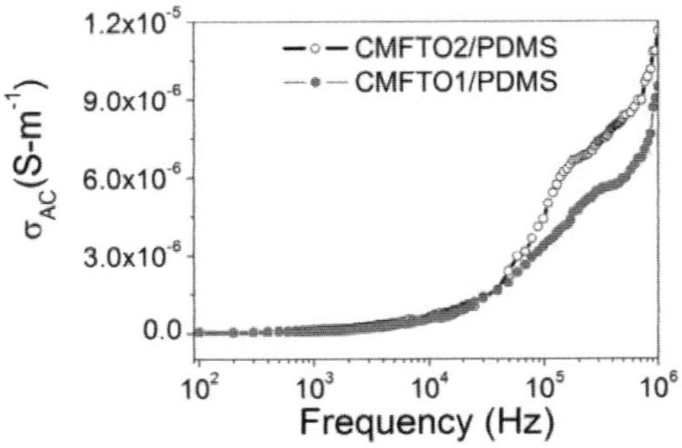

Figure 6. Frequency dependent AC conductivity (σ_{ac}) of CMFTO1/PDMS and CMFTO2/PDMS composite film.

The observed characteristics are a clear evidence of the transition of composite from an insulator to semiconductor zone. These effects in the crystalline oxide based materials are mainly developed due to the hopping of ions, which has greater possibility in both the flexible composite films [8].

The complex impedance spectra (Z′ vs. Z″) for both the CMFTO/PDMS composite films in the frequency range of 10^2-10^6 Hz are depicted in figure 7. The complex impedance plot of both the CMFTO/PDMS composite films show identical behavior. In the plot of figure 7, it could be observed that, the response curve diverges significantly from that of one impedance semi-circular arc. It could be noticed that the complex impedance spectra of both the composite film initiated to form a semicircle at 1.1×10^3 Hz before it diverged. These characteristics may be due to the conduction phenomenon, which happens due to the mobility of free electrons in the CMFTO/PDMS composite films.

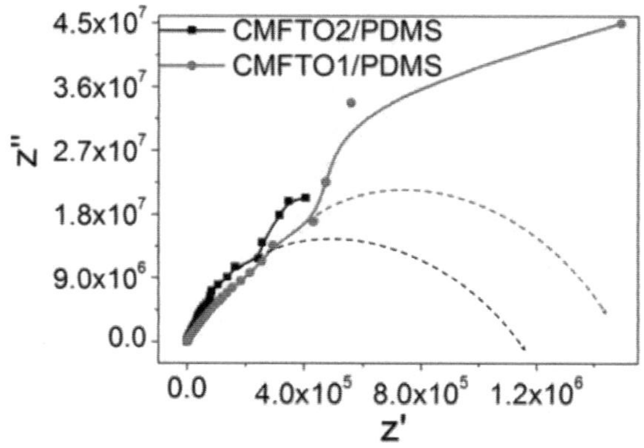

Figure 7. Nyquist plots of CMFTO1/PDMS and CMFTO2/PDMS composite films.

3.3. Mechanical properties

Static tensile characteristics of both the CMFTO/PDMS composite films and PDMS are presented in figure 8. Flexibility of CMFTO/PDMS composite films was evaluated by comparing with PDMS film. The Young's modulus of the composite films was evaluated in the strain range of 15-45%. The flexibility of both the composite films was tested by elongating the thin film up to 90%. The Young's modulus of CMFTO1/PDMS and CMFTO2/PDMS composite films was 0.53±0.49 and 0.80±0.21 MPa, respectively. Both the film shows extensively larger modulus value than the PDMS film, which is 0.27±0.08 MPa. The 90% elongation of both the composite film, suggest identical flexibility characteristic with PDMS film. Significantly lower Young's modulus of the CMFTO2/PDMS composite film than the PDMS film, confirms its excellent flexibility characteristics [9].

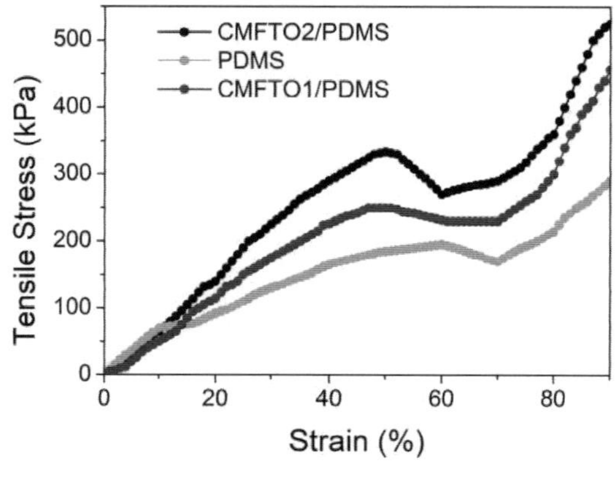

Figure 8. Static tensile properties of PDMS, CMFTO1/PDMS and CMFTO2/PDMS composite films.

4. Conclusions

In the current study, a new polymer ceramic composite was developed which shows higher dielectric constant (CMFTO2 ~350 & CMFTO1 ~150) than the previous results reported in the literature. The overall σ_{ac} of both the composite films was in increasing trend with increasing in applied alternating frequency, which indicates that dielectric characteristics of composite could be changed by the dynamic behavior of the polymer matrix. The dielectric constant of the newly developed CMFTO2/PDMS composite flexible film is significantly higher than results reported in the literature [10,11]. Therefore, the novel nanocomposite of CMFTO and PDMS composite film with trace amount of ferrous ferrite structures developed by solid-state step-sintering technique would be a potential dielectric material for several electronic miniaturization devices for capacitive memory, energy storage devices, actuators, sensor, switching and high frequency applications.

Acknowledgments

The authors are grateful to Prof. Noor Azuan Abu Osman and Dr. Sumit Pramanik for their motivation and support.

References

[1] Ramajo, L. A., Reboredo, M. M., & Castro, M. S. (2007). Characterisation of epoxy/BaTiO3 composites processed by dipping for integral capacitor films (ICF). Journal of Materials Science, 42(10), 3685-3691.

[2] Dias, C. J., Igreja, R., Marat-Mendes, R., Inacio, P., Marat-Mendes, J. N., & Das-Gupta, D. K. (2004). Recent advances in ceramic-polymer composite electrets. IEEE transactions on dielectrics and electrical insulation, 11(1), 35-40.

[3] Tripathy, A., Pramanik, S., Manna, A., Azrin Shah, N. F., Shasmin, H. N., Radzi, Z., & Abu Osman, N. A. (2016). Synthesis and characterizations of novel Ca-Mg-Ti-Fe-oxides based ceramic nanocrystals and flexible film of polydimethylsiloxane composite with improved mechanical and dielectric properties for sensors. Sensors, 16(3), 292.

[4] Zhang, L., Shan, X., Bass, P., Tong, Y., Rolin, T.D., Hill, C.W., Brewer, J.C., Tucker, D.S. and Cheng, Z.Y. (2016). Process and microstructure to achieve ultra-high dielectric constant in ceramic-polymer composites. Scientific reports, 6, 35763.

[5] Popielarz, R., Chiang, C. K., Nozaki, R., & Obrzut, J. (2001). Dielectric properties of polymer/ferroelectric ceramic composites from 100 Hz to 10 GHz. Macromolecules, 34(17), 5910-5915.

[6] Kumar, A., Singh, B. P., Choudhary, R. N. P., & Thakur, A. K. (2006). Characterization of electrical properties of Pb-modified BaSnO 3 using impedance spectroscopy. Materials chemistry and physics, 99(1), 150-159.

[7] Du, P., Lin, X., & Zhang, X. (2011, June). Dielectric constants of PDMS nanocomposites using conducting polymer nanowires. In Solid-State Sensors, Actuators and Microsystems Conference (TRANSDUCERS), 2011 16th IEEE International (pp. 645-648).

[8] George, M., Nair, S. S., John, A. M., Joy, P. A., & Anantharaman, M. R. (2006). Structural, magnetic and electrical properties of the sol-gel prepared Li0. 5Fe2. 5O4 fine particles. Journal of Physics D: Applied Physics, 39(5), 900.

[9] Ataollahi, F., Pramanik, S., Moradi, A., Dalilottojari, A., Pingguan-Murphy, B., Abas, W., Bakar, W.A., Osman, A. and Azuan, N. (2015). Endothelial cell responses in terms of adhesion, proliferation, and morphology to stiffness of polydimethylsiloxane elastomer substrates. Journal of biomedical materials research Part A, 103(7), 2203-2213.

[10] Nayak, S., Sahoo, B., Chaki, T. K., & Khastgir, D. (2013). Development of polyurethane–titania nanocomposites as dielectric and piezoelectric material. RSC Advances, 3(8), 2620-2631.

[11] Gan, W. C., & Majid, W. A. (2014). Effect of TiO2 on enhanced pyroelectric activity of PVDF composite. Smart Materials and Structures, 23(4), 045026.

Evaluation of surface integrity of WEDM processed inconel 718 for jet engine application

Priyaranjan Sharma[1], Ashis Tripathy[2] and Narayan Sahoo[3]

[1]Department of Mechanical Engineering, JK Lakshmipat University, Jaipur, India.
[2]Department of Electronics and Communication Engineering, JK Lakshmipat University, Jaipur, India.
[3]Department of Electrical Engineering, JK Lakshmipat University, Jaipur, India.

[*]Corresponding author: priya333ranjan@hotmail.com

Abstract. A unique superalloy, Inconel 718 has been serving for aerospace industries since last two decades. Due to its attractive properties such as high strength at elevated temperature, improved corrosion and oxidation resistance, it is widely employed in the manufacturing of jet engine components. These components require complex shape without affecting the parent material properties. Traditional machining methods seem to be ineffective to fulfil the demand of aircraft industries. Therefore, an advanced feature of wire electrical discharge machining (WEDM) has been utilized to improve the surface features of the jet engine components. With the help of trim-offset technology, it became possible to achieve considerable amount of residual stresses, lower peak to valley height, reduced density of craters and micro globules, minimum hardness alteration and negligible recast layer formation.

1. Introduction

Since past few years, aircraft manufacturing industries are continuously demanding complex shaped components for their utilization in jet engines in order to improve their efficiency and service life. Turbine section is amongst the most critical part in jet engine as it is exposed to a high temperature and high pressure gas stream [1]. Nickel based superalloys are an exceptional class of structural materials as they exhibit outstanding properties such as high mechanical strength at elevated temperature, excellent oxidation/corrosion resistance, hot hardness and high fatigue endurance limit [2]. For an example, Inconel 718 alloy comprises of about 34% of the total weight of aircraft engine [3] as shown in figure 1. An improved thermal, mechanical and chemical properties of Inconel 718 make it difficult to machine using traditional machine tool as it offers poor surface quality on the machined component and exhibits low process capability [4]. To overcome the issues of traditional machine tool, some advanced machine tools are successfully implemented, which uses energy particles such as laser (photon), electron and ions for removal of material. Wire electrical discharge machining (WEDM) is most widely accepted advanced machine tool due its capability in manufacturing of complex shape component with the help of CNC programming. Moreover, WEDM process offers minimum surface damage to the workpiece as it utilizes the ultrafine electrical discharge and produces the surface features within the tolerance band of 1–0.1 μm. Previous literature [5-8] indicates that a significant research was conducted on the machinability enhancement of Inconel 718 superalloy for various engineering applications. Some researchers [9, 10] have utilized the statistical tool and evolutionary algorithm to determine the feasible range of WEDM control parameters for improved machining performance.

Despite the abundant literature available in the field of WEDM process, very few studies have been reported on evaluation of surface integrity of aircraft components madeup of Inconel 718 using WEDM process. In the current study, special features of WEDM process such as trim-offset technology was utilized to achieve the feasibility in manufacturing of jet engine components. The adopted technology also offers less surface damage to the machine components and helps to obtain the better surface integrity in terms of minimum hardness alteration, negligible recast layer formation, low density of craters, micro globules creation and, also produces minimum residual stresses.

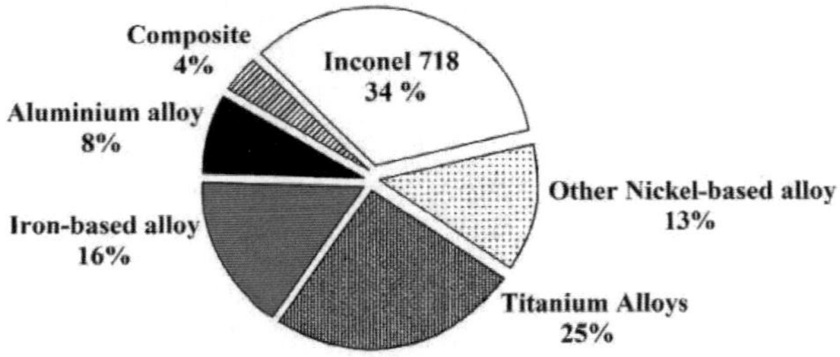

Figure 1. Material distribution in GE CF6 aircraft engine [2].

2. Materials and Methods

2.1 Material selection and preparation

Inconel 718 plate (250 mm and 100 mm × 10 mm) was used as experimental workpiece. The alloy was procured from 'Special Metal' (India). The chemical composition of Inconel 718 has been shown in table 1. Further, the workpiece was heat treated in a tubular furnace at a temperature of 800 °C to avoid the chances of distortion during manufacturing of aircraft component. Before actual experimentation, it was confirmed that workpiece is free from dust and rust to ensure the proper functioning of WEDM during edge finding.

Table 1. Chemical composition of as-received Inconel 718.

Alloy (%)		Ni+Co	Cr	Fe	Nb+Ta	Mo	Ti	Al	Co	Cu	C	Mg	Si
Inconel	Min.	50	17	Bal	4.75	2.8	0.65	0.2					
706	Max.	55	21		5.50	3.3	1.15	0.8	1	0.3	0.08	0.35	0.35

2.2. Experimental details

The experimental work was performed on WED machine (Model: ECOCUT from Electronica Machine Tools, Pune, India). Zinc coated wire was selected as tool electrode whereas de-ionized water was used as dielectric fluid. An optimum value of wire feed (6 m/min) and flushing pressure (1.96 bar) was selected based on preliminary investigation [11]. In the current study, rough cut technology was used for basic cutting of complex shape profile through aircraft components, however trim-offset technology was employed to obtain the high dimensional accuracy and better surface integrity on the aircraft component. Six control parameters were selected for manufacturing of aircraft components as shown in table 2. The feasible range of control parameters was estimated based on the several trial experiments. Within the range of parameters selected, no gap short and wire breakage was reported. CNC programming of complex aircraft components was carried out using 'ELCAM' software. The mechanism of WEDM operation has been shown in figure 2. Initially, a high potential difference was applied between the wire electrode and workpiece using a pulse generator which produces a strong electric field. If the bonding strength of electrons is less than the electric field strength, then electrons are cold emitted from wire electrode and, attracted towards the workpiece material. Thus, electrons gain some kinetic energy and collide with the molecules of dielectric fluid resulting in ionization. Due to ionization of dielectric fluid, plasma channel forms which has very low electrical resistance. Therefore, more number of electrons and ions are formed between wire electrode and workpiece due to ionization. All of a sudden, a large density of electron starts flowing from wire electrode to the workpiece and ions from workpiece to wire electrode. These avalanche motion of electrons and ions are visually seen as spark. Thus, electrical energy is converted into thermal energy of the spark. Generally, electrical discharge has a temperature of around 10,000°C which is more than enough to melt and vaporize the any conductive material.

Table 2. Technology table for manufacturing of jet engine components.

WEDM Technology	Pulse current (A)	Pulse on Time (μs)	Pulse off Time (μs)	Servo Voltage (V)	Servo Feed (mu)	Wire offset (μm)
Rough-cut	12	112	36	20	2150	0
Trim-offset	2	105	20	10	200	50

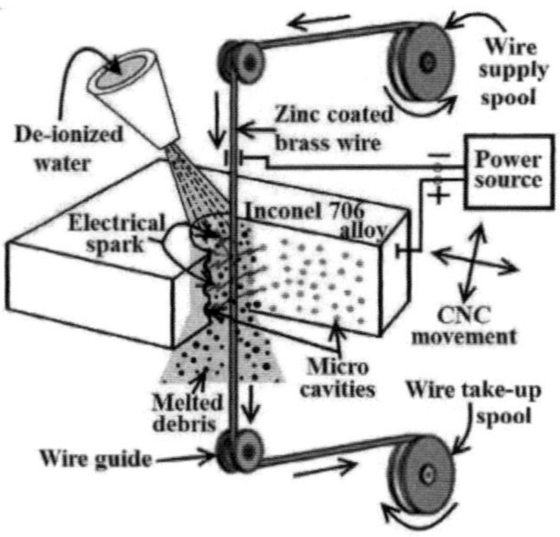

Figure 2. Principle of WEDM operation [10].

3. Result and Discussion

3.1 Surface Topography Analysis

For surface topography analysis, 3D laser microscope was utilized which is capable enough to measure the micro/nano level surface texture due to its low spot diameter of 0.4 μm. A scan area of 106 μm ×106 μm was selected for topographic investigation. The surface topography result was shown in figure 3 which shows that rough cut surface has higher peak to valley height (Rz) whereas trim cut surface exhibits the lower Rz value. The behavior could be explained by the fact that in the rough cut mode, pulse on time is too high which increases the spark intensity. Therefore, consequently, more amount of material is melted from the machined surface. A part of the molten material was flushed away by pressurized wave generated in the absence of plasma channel. When remaining molten metal re-solidify, forms a large crater on the machined surface. Therefore, machined surface shows the high Rz value of 23 μm as shown in figure 3(a). However, in trim cut mode, ultrafine electrical discharges are produced which melts only a small amount of material from the machined surface which can be easily flushed away by pressurized wave. Thus, forms micro/nano craters on the machined surface and shows the low Rz value of 11 μm as shown in figure 3(b).

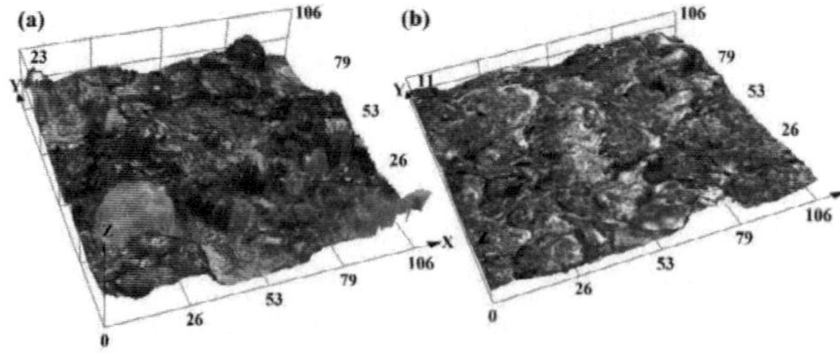

Figure 3. Surface topography analysis of WED machined surface, (a) Rough cut surface; (b) Trim cut surface.

IOP Conf. Series: Materials Science and Engineering **323** (2018) 012019 doi:10.1088/1757-899X/323/1/012019

3.2. Microstructure Analysis
For microscopic investigation, machined surface was exposed to scanning electron microscope (SEM). The SEM analysis revealed that under rough cut mode, there is a high density of micro globules, micro holes, craters and melted debris on the machined surface as shown in figure 4(a). This is because, under rough cut, discharge pulse has a high pulse on time which increases the spark intensity. This, in turn, melts more amount of material and forms larger crater on the machined surface. A part of molten material is flushed away by pressurized waves, however some air bubbles get entrapped in it. When remaining molten material re-solidify, these air bubbles collapse and generate micro holes on the machined surface. On the other hand, the propensity of micro globule and crater formation are considerably reduced in trim cut mode due to ultrafine electrical discharge except few micro holes and melted debris. These ultrafine discharge pulses melt only a small amount of material from the machined surface which can be easily flushed away by pressurized wave and allow the formation of micro/nano cavities on the machined surface as shown in figure 4(b).

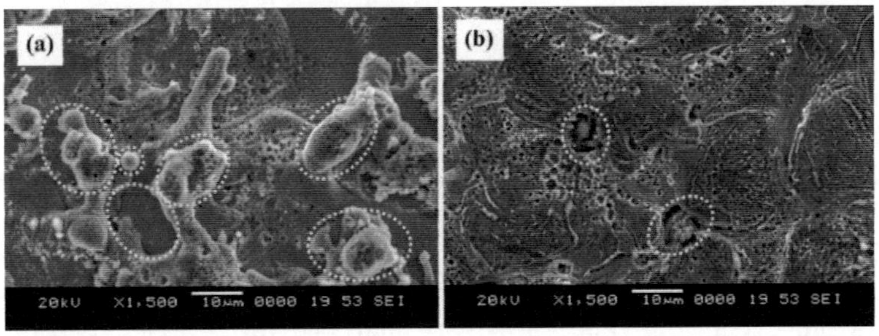

Figure 4. Micrograph of WED machined surface, (a) Rough cut surface; (b) Trim cut surface.

3.3. Recast Layer Analysis
Recast layer is a layer which is commonly observed on the machined surface due to melting and resolidification of material. To measure the recast layer thickness, sample are cross-sectionally polished at a rotational speed of 300 rpm using a progressive grade of silicon carbide (SiC) papers and diamond paste. The recast layer formed on the machined components has been shown in figure 5 which indicates significant thermal damage in case of rough cut mode however minimum surface damage was detected under trim cut mode. The behavior could be explained by the fact that under rough cut mode, more thermal energy has been transferred to the workpiece material which melts consequently more amounts of material from the workpiece. A part of the molten material was flushed away by pressurized waves when remaining molten material re-solidify and form a thick recast layer on the machined component as shown in figure 5(a). But, recast layer formation has a detrimental effect on aircraft component. With the use of ultrafine discharge pulse in trim-offset technology, it became possible to reduce the recast layer thickness significantly. These ultrafine discharge pulses melt only a small amount of material which can be easily flushed away by pressurized waves generated in the absence of plasma channel. Hence, recast layer formation significantly reduced in trim cut mode.

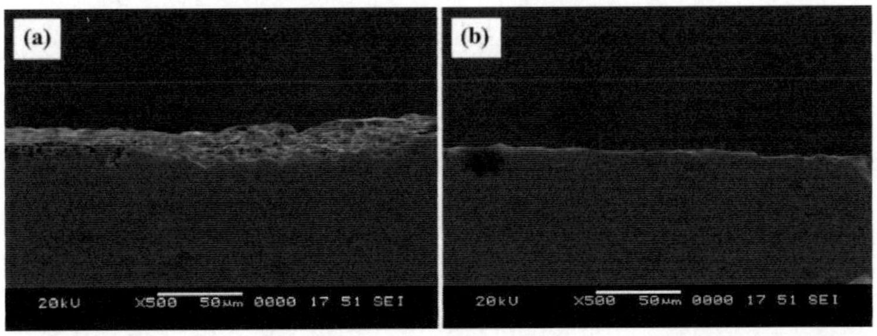

Figure 5. Recast layer formed on WED machined surface, (a) Rough cut surface; (b) Trim cut surface.

3.4. Microhardness Analysis

To measure the subsurface microhardness, sample are cross-sectionally cold mounted using acrylic resin. Then, mechanical polishing was done using sequential grade of SiC papers and diamond paste in order to obtain the mirror finish structure. Microhardness profile of WEDM processed Inconel 718 has been shown in figure 6 which indicates that microhardness of WED machined surface has been decreased to a depth of 80 μm below the machined surface under rough cut. However, no significant change in microhardness was detected under trim cut mode. The behavior could be explained by the fact that Inconel 718 has very low carbon content which won't make the surface harder even quench by dielectric fluid.

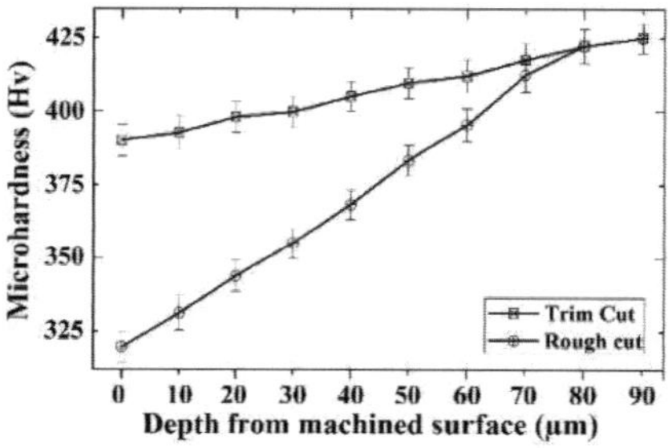

Figure 6. Microhardness analysis of WED machined surface.

3.5. Residual Stress Analysis

The residual stresses induced in WED machined Inconel 718 was measured using 'PROTO-iXRD' stress measuring system. Due to sudden heating and cooling during WEDM process, high temperature gradient is acting on the machined component. When, workpiece material was melted due to the thermal energy of the spark. A part of the molten material was flushed away by pressurized wave when remaining molten material re-solidify then its shrinkage was opposed by the bulk material. Hence, recast material exhibits the tensile residual stresses whereas bulk material exhibits the compressive residual stresses as shown in figure 7. For jet engine application, tensile residual stress should not exceed beyond 850 MPa as it produces harmful effect on aircraft component and significantly reduces the service life. In the current study, residual stresses were significantly reduced under the trim cut due to negligible recast layer formation and make the WEDM process highly suitable for manufacturing of jet engine components.

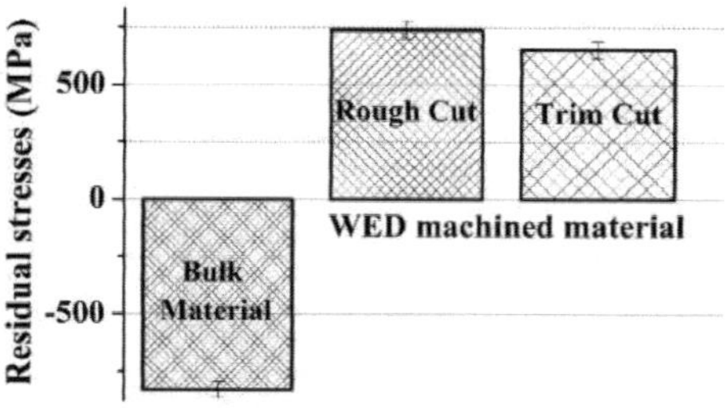

Figure 7. Residual stresses generated in aircraft component during WEDM process.

4. Conclusion

Based on the experimental investigation, following conclusion were derived:

- The surface topography analysis exposes that WEDM trim-offset technology offers lower Rz value of 11 μm and significantly improve the profile accuracy.
- Microstructure analysis reveals that trim-offset technology offers no micro globules and no crater formation on machined components except few micro holes and melted debris.

- The recast layer analysis exposes that trim-offset technology offers almost negligible recast layer thickness due to ultrafine discharge pulse and make it highly suitable for manufacturing of jet engine components.
- Microhardness analysis reveals that subsurface microhardness has been decreased to a depth of 80 μm below the machined surface under rough cut mode. However, no significant change in subsurface microhardness was detected under trim cut mode.

Residual stress analysis discloses that machined surface exhibit tensile residual stresses due to sudden heating and cooling during WEDM process. However, trim-offset technology offers very low level of tensile residual stresses on the machine component.

Acknowledgement
The authors would like to thank the 'Central Manufacturing Technology Institute (CMTI)' and 'Advanced Machine Tool Testing Facility (AMTTF)' for their indebted support for testing the aircraft components. We would also like to thanks the Dr. D. Chakradhar and Prof. Narendranath S. for their valuable suggestion and motivation.

References

[1] Rolls Royce. (2015). *The jet engine*. John Wiley & Sons.
[2] Campbell FC (2006) Manufacturing technology for aerospace structural materials. Elsevier.
[3] Schafrik R and Sprague R (2004) Saga of Gas Turbine Materials: Part III. *Adv Mater Processes*, 29–33.
[4] Choudhury IA, El-Baradie MA (1998) Machinability of nickelbase super alloys: a general review. *J Mater Process Technol* 77 (1):278–84.
[5] Ahmad, S., & Lajis, M. A. (2013). Electrical discharge machining (EDM) of Inconel 718 by using copper electrode at higher peak current and pulse duration. In *IOP Conference Series: Materials Science and Engineering* (Vol. **50**, No. 1, p. 012062). IOP Publishing.
[6] Li, L., Li, Z. Y., Wei, X. T., & Cheng, X. (2015). Machining characteristics of Inconel 718 by sinking-EDM and wire-EDM. *Materials and Manufacturing Processes*, *30* (8), 968-73.
[7] Klocke F, Welling D, Klink A, Veselovac D, Nothe T, Perez ¨ R (2014) Evaluation of advanced wire-EDM capabilities for the manufacture of fir tree slots in Inconel 718. *Procedia CIRP* 14:430–35
[8] Aspinwall DK, Soo SL, Berrisford AE, Walder G (2008) Workpiece surface roughness and integrity after WEDM of Ti–6Al–4V and Inconel 718 using minimum damage generator technology *CIRP Ann Manuf Technol* 57(1):187–90.
[9] Saha, A., & Mondal, S. C. (2016). Multi-objective optimization in WEDM process of nanostructured hardfacing materials through hybrid techniques. *Measurement*, *94*, 46-59.
[10] Sharma, P., Chakradhar, D., & Narendranath, S. (2017). Analysis and Optimization of WEDM Performance Characteristics of Inconel 706 for Aerospace Application. *Silicon*, 1-10.
[11] Sharma, P., Chakradhar, D., & Narendranath, S. (2015). Evaluation of WEDM performance characteristics of Inconel 706 for turbine disk application. *Materials & Design*, *88*, 558-66.

IOP Conf. Series: Materials Science and Engineering **323** (2018) 012020 doi:10.1088/1757-899X/323/1/012020

Single axis control of ball position in magnetic levitation system using fuzzy logic control

Narayan Sahoo[1], Ashis Tripathy[2] and Priyaranjan Sharma[3]

[1]Department of Electrical Engineering, JK Lakshmipat University, Jaipur, India
[2]Department of Electronics and Communication Engineering, JK Lakshmipat University, Jaipur, India
[3]Department of Mechanical Engineering, JK Lakshmipat University, Jaipur, Rajasthan, India.

*Corresponding author: narayansahoo@jklu.edu.in

Abstract. This paper presents the design and real time implementation of Fuzzy logic control(FLC) for the control of the position of a ferromagnetic ball by manipulating the current flowing in an electromagnet that changes the magnetic field acting on the ball. This system is highly nonlinear and open loop unstable. Many un-measurable disturbances are also acting on the system, making the control of it highly complex but interesting for any researcher in control system domain. First the system is modelled using the fundamental laws, which gives a nonlinear equation. The nonlinear model is then linearized at an operating point. Fuzzy logic controller is designed after studying the system in closed loop under PID control action. The controller is then implemented in real time using Simulink real time environment. The controller is tuned manually to get a stable and robust performance. The set point tracking performance of FLC and PID controllers were compared and analyzed.

1. Introduction

Magnetic levitation is a technique in which gravitational acceleration is negated which in turn leads to free suspension of objects without any support of mechanical system. Since magnetic levitation provides a near frictionless environment, this particular trait makes it a popular choice in many research fields, such as high speed transportation, mechanical engineering, automatic control and biomedical engineering [1-5]. The maglves however are inherently nonlinear and open loop unstable in nature. Therefore, in order to achieve desired performance such as excellent setpoint tracking, minimum overshoot and minimum steady sate error, stabilizing feedback controllers are required which are based on a linearized model. Since there could be many disturbances acting on the operational maglev system and the performance of linearized model can be affected when it deviates from the linearized point, classical control approach is not effective, therefore an intelligent control technique is demanded for handling disturbance rejection and parameter variations [6-8].

The current research article encompass the design of a fuzzy logic controller (FLC) using fuzzy logic toolbox and then implementation of the FLC in real time environment for the magnetic levitation system using SIMULINK. The performance of FLC is compared with the existing PID controller. The remainder of the paper is structured as follows. Section 2, section3 represents the modeling of the system and design of FLC respectively. While, section 4 represents the experimental analysis of the present research work which shows the comparative performances of FLC and PID. The last section of the paper i.e. section 5, presents the conglomeration of the whole research work.

2. Mathematical Modelling of Maglev

Figure 1 gives a pictorial view of Laboratory setup of magnetic levitation system, also including the auxiliary components of the system such as IR Transmitter & receiver for measurement of ball position, the electromagnet which acts as actuator, the iron ball and the data acquisition card connected to a personal computer provided by Feedback instruments(33-210).

Figure 1. Schematic illustration of Feedback instruments (Maglev(33-210).

2.1. Characteristics of magnetic levitation system

The Maglev device consists of three important components: an infrared (IR) sensor, an electromagnet which acts as an actuator and a feedback controller which operates around a linearized point from the tip of the electromagnet. The physical parameter values of Maglev (33-210) shown in figure 1 are as follows: the iron ball used has a mass of 20 g and the electromagnet coil has a resistance of 22 Ω, inductance of 227 mH at 1kHz .The core diameter, coil diameter are 25 mm and 80 mm and having 2850 number of turns. The mean crossectional area of the coil is 21.5 cm^2 and the leangth of the coil is 65 mm.

In the linearized zone the photo detector of the Maglev gives a voltage Z which is proportional to the amount of IR light it senses, has a relation with the distance X of the levitating object(a ferromagnetic ball or iron ball) from the tip of the electromagnet can be expressed as [9]

$$Z = \beta X + \overline{Z} \tag{1}$$

Where $\beta > 0,\ \overline{Z}$ are constants such that $Z \in (-1.5\,V, 1.5\,V)$

The current I in the electromagnet is varied by an inner control loop and have a liner relationship with input voltage E as

$$I = \sigma E + \overline{I},\ \sigma > 0 \tag{2}$$

Where constant $\overline{I}$ is the magnitude of current that is necessary for keeping $Z = \overline{Z}$.

A deviation from the equilibrium point is shown by using lower case, hence equations (1) and (2) can be rewritten as

$$z = \beta x \tag{3}$$

$$i = \sigma e \tag{4}$$

The nonlinear differential motion equation which governs the single axis movement is given by

$$m\frac{d^2X}{dt^2} = mg - p\frac{I^2}{X^2} \tag{5}$$

Where $p > 0$ is an electromechanical conversion coefficient, m is the mass of the sphere, g is the acceleration of gravity, and I is the current fed to the coil. With I kept constant, the equilibrium distance from the tip of the electromagnet is expressed as

$$X_o = I_o\sqrt{\frac{p}{mg}} \tag{6}$$

This nonlinear dynamic equation of the MAGLEV is linearized around the equilibrium position X_o using first order Taylor's expansion gives rise to the following expressions.

$$m\frac{d^2x}{dt^2} = \lambda x - \tau i \tag{7}$$

where,

$$\tau = \frac{2pI_o}{{X_o}^2} = \frac{2\sqrt{pmg}}{X_o} \tag{8}$$

$$\frac{d^2z}{dt^2} = \omega z - De \tag{9}$$

where $\omega = \frac{2g}{X_o}$ and $D = \frac{2\sigma\beta}{X_o}\sqrt{\frac{pg}{m}}$

The resulting transfer function after linearization around equilibrium point is given by

$$G(s) = \frac{L[z]}{L[e]} = \frac{-D}{s^2 - w} \tag{10}$$

As seen from equation (10) the system has a pole in the right hand side of the S-plane, therefore the system is open loop unstable, hence we need to develop a suitable stabilizing controller. The parameters used for modelling and its values for magnetic levitation system is as follows height to voltage multiplier (B) is given as 143.48, the electromechanical constant (p) has a value of 2.501779922×10^{-5} . The height to voltage constant ($\overline{Z}$) and sensor gain (σ) has a value of -2.8 and 1.045 respectivly and g reperesents acceleration due to gravity havng a vlue of 9.8 m/s^2. Where E, I and X represents controller output, current fed to the coil and output height respectively. The next section describes how to design and implement a fuzzy logic controller for the magnetic levitation system.

3. Design of Fuzzy Logic Controller (FLC)

As discussed in the introductory section, magnetic levitation systems are naturally nonlinear and unstable. The nonlinear magnetic levitation system is first linearized around an operating point and then position control is done by employing a PID controller. Since the stability of nonlinear system degrades when initial condition deviates from the equilibrium zone and to achieve a large stable region with desired performance specification application of fuzzy logic for a robust controller design is reasonable[10]. The next section describes the implementation of fuzzy logic controller.

3.1. Implementation of fuzzy logic controller (FLC) for magnetic levitation system

In this research work, a two input fuzzy controller is designed and applied to a laboratory-scale magnetic levitation system (Feedback, 33-210). FLC uses the error and rate of change of error for calculating the controller output. The goal of using FLC is to minimize the error. The error here is the difference between reference position and available position of the steel ball. The output of fuzzy controller is created by rules which are composed of these two inputs with linguistic definitions of the system. The general rule is that if error value is negative and change of error value is negative then output of controller is negative or if the error value is positive and change of error value is positive then the output of controller will be positive.

Figure 2 represents the two inputs error and change in error to the FLC. The error and change in error have seven membership functions, five triangular and two trapezoidal: negative big (NB), negative medium (NM), negative small (NS), zero (ZE), positive small (PS), positive medium (PM), positive small (PS) in order to get desired value with high accuracy. The range of these two inputs are taken to be [1 -1].

Figure 3 represents the output of the FLC. The output is also having seven membership functions five triangular and two trapezoidal: negative big (NB), negative medium (NM), negative small (NS), zero (ZE), positive small (PS), positive medium (PM), positive small (PS) in order to get desired value with high accuracy. The range of this input is taken to be [2 -2].

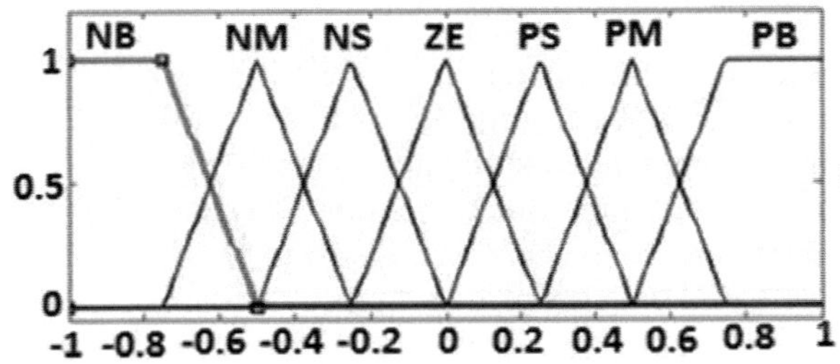

Figure2 . Linguistic variable Error/ Change in error.

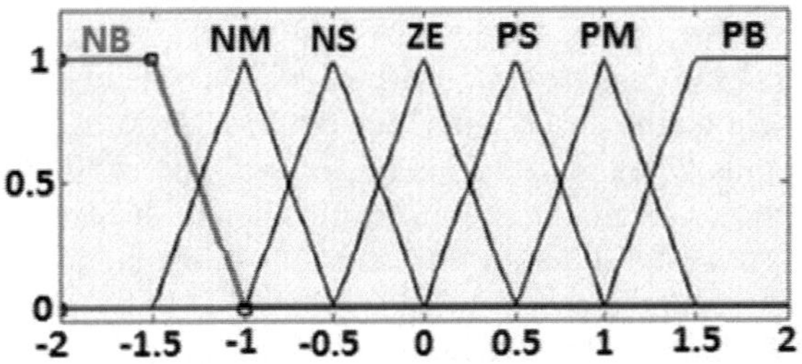

Figure 3. Linguistic variable output.

3.2. Fuzzy rules

The general rules for position control of steel ball in magnetic levitation system is that if ball position is below the reference position then increase the voltage across the electromagnetic coil (control voltage) hence current flowing in the coil will also increase and if ball position is above the reference position then bring it down to reference position by decreasing the control voltage. In the process of producing output voltage with fuzzy controller the position error must be minimized. The bigger position error causes the bigger controller input. So two inputs make 49 rules in order to get accurate value and efficiency. Table 1 represents the fuzzy rule base used for control of the ball position in the magnetic levitation system.

Table 1. Rule base.

Error \ Change in error	NB	NM	NS	ZE	PS	PM	PB
NB	NB	NB	NB	NB	NM	NS	ZE
NM	NB	NB	NB	NM	NS	Z	PS
NS	NB	NB	NM	NS	ZE	PS	PM
ZE	NB	NM	NS	ZE	PS	PM	PB
PS	NM	NS	ZE	PS	PM	PB	PB
PM	NS	ZE	PS	PM	PB	PB	PB
PB	ZE	PS	PM	PB	PB	PB	PB

3.3. Scaling factors

The scaling factors presented in table 2 for error, change in error and controller output was used in order to scale the desired inputs and outputs within the range of membership function's range. They are calculatedon a trial and error basis.

Table2. Tuning Parameters.

Tuning parameters	Scaling factors
Error	0.22
Change in error	0.01
Controller output	-10

3.4. Hardware implementation

The real time analysis of the controller is performed in the magnetic levitation system supplied by Feedback Instruments, UK (33-210). A data acquisition system with inbuilt (Digital to analog converter) DAC and (analog to digital converter) ADC is used for communication between the controller (computer) and MLS. Figure 1 presents the pictorial view of FLC connected with maglev (33-210) when the device was operated in real time.

IOP Conf. Series: Materials Science and Engineering **323** (2018) 012020 doi:10.1088/1757-899X/323/1/012020

4. Performance Analysis of Controllers

The FLC parameters are tuned with trial and hit method. The criteria followed while tuning was to give a stable steady state performance with minimum error.The performance of FLC is analysed for various types of setpoints in the presence of many un-measureable disturbances affecting the process such as air circulation, moisture, change in coil resistance with time due to the generated heat in coil, magnetic field distortion due to other magnetic elements etc. Step input was given to the system for analysis and the performance of designed FLC was compared with the professionally tuned PID.

Figure 4 represents the step response of the maglev (33-210), for step input (Magnitude changing from 12.5 mm to 9 mm at t = 35 s); it was found that FLC is giving an over-damped performance with constant steady state error. Table 3 gives the time response specifications of the system when step input was applied, there is a step change at t = 35 s as seen from the table 3 PID have a high overshoot in comparison to fuzzy logic controller (FLC) which is giving an over damped response at the same time the FLC have a shorter settling when compared to PID controller. The error analysis was also done for both the controllers as shown in table3.

Table 3. Performance analysis of controllers for step change in setpoint.

Performance Criteria	Controller	
	PID	FLC
% Overshoot	40	0
Rise time - 0% to 100% (sec)	0.0418	0.135
Settling Time (sec)	1.16	0.4
% Steady state error	0	28.57
ISE of MLS output (10^{-6})	192	627
ISE of controller output	10708.7	139.8

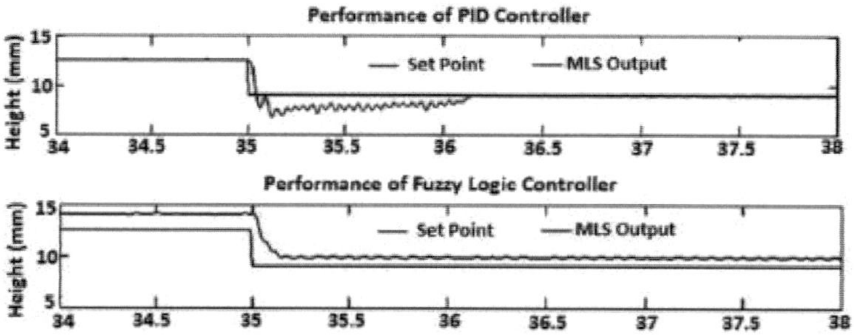

Figure 4. Step change response of magnetic levitation system.

Figure 5 represents the controller output of both the PID and FLC as shown in the figure FLC found to put maximum effort by continuous change in controller output. For PID; when rate of change of setpoint is high, output shows derivative kick.

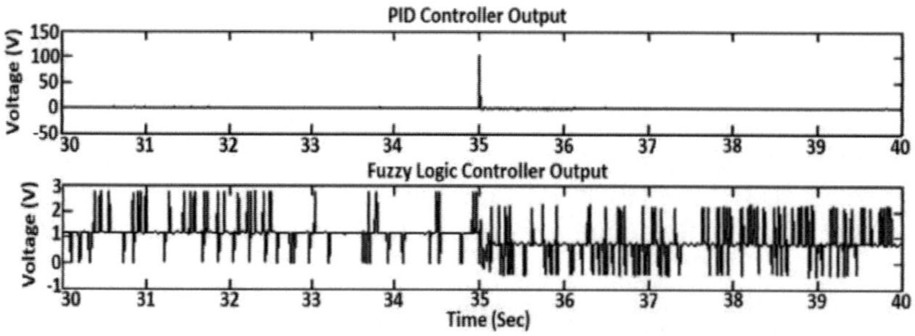

Figure 5. Controller output for Step change in setpoint.

During the experiment it was observed that there is a huge scope in improving the performance of both FLC by using better tuning methods and by incorporating further intelligence through Neural Networks and Heuristic algorithms in the controller.

5. Conclusions

The paper gives theoretical description as well as the real time implantation about the design of Fuzzy controller for a laboratory scale magnetic levitation device.The nonlinear system is linearized about an operating point .The FLC was designed in the linearized zone. The performance of the controller was compared with already existing PID controller on the basis of time response analysis. After the comparison it is concluded that although the fuzzy logic controller shows constant steady state error but the FLC have better performance in terms of lower settling time and overshoot. The FLC takes only 0.4 sec to settle down as compared to 1.16 sec by PID controller also the fuzzy FLC has 0 % overshoot while the PID has a staggering 40 % overshoot. While the speed of response for both the controllers are nearly same. Error analysis was also done for both the controller output (Integral square error) ISE for FLC was found better than the PID. ISE for FLC is 139.8 while the ISE for the PID is at staggering magnitude of 10708.7. The scaling factors for the FLC which is equivalent to gain of the classical controllers are very important and are found by hit and trial method. The performances of FLC can be improved with certainty by proper tuning of membership functions and scaling factors in case of Fuzzy Logic controller.

Acknowledgments

The authors would like to express their gratefulness to the Control & Industrial automation Department, NIT Silchar for allowing to perform experiment on the MAGLEV (33-210) setup. The authors would also grateful to Prof. B.K Roy and Mr. Lalu Seban for the motivation to carry out the research.

References

[1] Yadav, Shekhar, S. K. Verma, and S. K. Nagar. "Optimized PID controller for magnetic levitation system." IFAC-PapersOnLine 49.1: 778-782, 2016.

[2] J. Delamare, J. Yonnet, and E. Rulliere. A compact magnetic suspension with only one axis control. IEEE Transactions on Magnetics, 30(6):4746–4748, 1994.

[3] T. Mizuno and R. Yoshitomi. Vibration isolation system using zeropower magnetic suspension. 1streport. principles and basic experiments. Transactions of the Japan Society of Mechanical Engineers. C, 68(673):2599–2604, 2002.

[4] A. Karimi, L. Miˇskoviˊc, and D. Bonvin. Iterative correlation-based controller tuning with application to a magnetic suspension system. Control Engineering Practice, 11(9):1069–1078, 2003.

[5] K. Jung and Y. Baek. Precision stage using a noncontact planar actuator based on magnetic suspension technology. Mechatronics, 13(8-9):981–999, 2003.

[6] P. Yang, Q. Zhang, and L. Li. Design of fuzzy weight controller in single-axis magnetic suspension system. In Intelligent Control and Automation, 2002. Proceedings of the 4th World Congress on, volume 4, 2002.

[7] W Hurley, M . Hynes, and W Wolfle. PWM control of a magnetic suspension system. IEEE Transactions on Education, 47(2):165–173, 2004.

[8] C. Kuo, T. Li, and N. Guo. Design of a novel fuzzy sliding-mode control for magnetic ball levitation system. Journal of Intelligent and Robotic Systems, 42(3):295–316, 2005.

[9] Seban, Lalu, Narayan Sahoo, and B. K. Roy. "Multiple model based predictive control of magnetic levitation system." India Conference (INDICON), 2014 Annual IEEE. IEEE, 2014.

[10] S. Lei, A. Palazzolo, and A. Kascak. Fuzzy logic intelligent control system of magnetic bearings. In IEEE International Fuzzy Systems Conference, 2007. FUZZ-IEEE 2007, pages 1–6, 2007.

IOP Conf. Series: Materials Science and Engineering **323** (2018) 012021 doi:10.1088/1757-899X/323/1/012021

Experimental investigation of springback in air bending process

Aysha Alhammadi, Hafsa Rafique, Meera Alkaabi and Jaber Abu Qudeiri

Faculty of Engineering, United Arab Emirate University, Al Ain, UAE.

*Corresponding author: jqudeiri@uaeu.ac.ae

Abstract. Bending processes is one of the important processes in sheet metal forming. One of the challenge that faces the air bending process is springback, which happens due to the elastic recovery during unloading stage. An accurate analysis of springback during the bending process is crucial to achieve a required bend angle. This paper will investigate the springback experimentally by changing many parameters such as tested material, die opening, thickness, etc. and finding its effect on the value of springback. Additionally, the paper will investigate the effect of loading time at the end of loading stage on the springback by proposing a multistage bending technique (MBT). In MBT, the loading will stop during loading stage just before the end of this stage and it will restart again shortly after. In this study, three sheet metals with different thickness will be examined, namely stainless steel, aluminium and brass. Artificial neural network (ANN) will be utilized to develop a prediction model to predict springback based on the experimental results.

1. Introduction

In order to achieve an accurate bending process, an analysis of springback in the bending process is crucial to determine. Many studies introduced a number of analytical models based on the material properties and the tool geometry that are available to predict springback. Many studies discussed the springback for different materials, for example, Senol, et al. [1] investigated the air bending process of stainless-steel sheet of different thicknesses and bend angles through Artificial Neural Network (ANN). Experimentation of the air bending processed is also carried out and the results are used in artificial neural network development to show the probability of ANN based on experimentation. Also, Jiang, et al. [2] conducted a study to investigate the coupling effects of the bending angle of a TA18 (titanium alloy) tube and the material properties on the springback angle during NC bending. The result showed that the influence of the material properties of the titanium alloy increases with the bending angle. Han et al. [3] proposed numerical and experimental methods to investigate effects of process parameters on springback characteristics of AZ31 magnesium alloy bent profile by doing the process of applying warm tension-rotation bending. The results indicated that the relationship between the forming temperature and the springback angle seemed to be linear. Heng et al. [4] experimented the effects of forming parameters on the springback behaviors in 6061-T4 Al-alloy tube bending, the results found that the springback becomes larger with increasing of the bending velocity, the tube-die clearance, the relative bending radius, the tube-pressure die friction and relative push assistant speed. Also, Wang et al. [5] investigated the springback and neutral layer for AZ31B magnesium alloy sheet by V-bending tests at temperatures from 50 to 300 °C. The results show that the neutral layer shifts to the tension zone of the sheets. The coefficient of neutral layer (k-value) decreases with the increase of temperature and punch radii. Also, Lawanwomg [6] used a new technology to eliminate springback of 980Y HSS sheets on U-bending process, where the bottom plate is additionally bent with a counter punch at the final stage of U-bending process. The result found were that the major mechanism of the springback elimination is 'spring-go' by releasing negative bending moment, which has been generated by bottom pushing-up. Wiriyakorn and Thipprakmas [7] examined a new spring-back factor for a wiping die bending process to achieve a more accurate predicted bend angle of aluminium A1050-H14 using the two-dimensional plane strain modelling of an elasto-plastic, finite element model. The result showed that the spring-back factor depended not only on the ratio of the die radius to workpiece thickness but also on the bend angle, which is the new springback factor. Gisario et al. [8] used an external-force laser assisted bending process of Grade 2 CP titanium and AA 7075 T6 aluminum sheets. The experiment showed that both metal sheets could be bent to high angles with very low sharp fillet radii and control of springback using appropriate combination of tuning the contact pressure and selective laser heating of the bending zone by irradiation with a high-power diode laser. Badr et al. [9] conducted a research in which they used Ti-6Al-4V which is a high strength titanium alloy at room temperature to apply and verify the accuracy of the model using bending processes such as V-die bending and the roll forming process and then verified it with experimental results. The model demonstrated that the springback is lower in roll forming compared to simple bending. Furthermore, Sofuoglu et al. [10] investigated the springback in the three-point bending operation of AA6082T6 tubes and formed a numerical model and the results were compared with experimental outputs to verify the results. This study specifically focuses on the springback of the three-point bend that is not used on tubes in the common industry. Gisario et al. [11] in their research used the external force laser-assisted bending of Titanium Grade-2 flat sheets to achieve sharp bending angles (>140°) with small fillet radii to analyse the influence of the operational parameters, on bending angles and fillet radii of the metal substrates. The experimental results showed that shaping of the substrates can be performed with great reliability, being springback largely minimised in broad operational ranges.

Many methods and techniques were introduced to estimate springback such as ANN [12]. For example, Inamdar et al. [13] used ANN to determine the backpropagation (BP) of error, where the ANN was subsequently trained with over 400 plane strain bending experimental data. The results showed that the accuracy of predictions depended more on the number of training patterns used than on the ANN architecture. Also, Song et al. [14] used high-strength Ti–3Al–2.5V tube as the model and studied the effects of numerical parameters on prediction accuracy and computation efficiency of springback simulation of high-strength Ti–3Al–2.5V tube with tube rotary-draw-bending. The results found were solid and continuum-shell elements predict the experimental results well; specifying damping factor typically works well in Abaqus/Emplicit simulation of springback. Song and Yu [15] used the three-point bending method for the T-section beam bending process. It used ANN artificial neural network to make a prediction model along with numerical simulations using finite element method to study the effect of material properties on springback. These application examples indicate that the proposed approach could achieve an allowable straightness error. Maati et al. [16] investigated the influence of constitutive modelling on the prediction of the degree of springback in the case of a stretch bending test on titanium sheets by means of numerical simulations. It also investigated the dependence of some process parameters such as the clamping force on springback. Furthermore, Yang et al. [17], conducted a study where they used an analytical model to predict the springback in air-bending of Advanced High-Strength steels (AHSS), considering the special properties of these materials. Their study also analyses the comparison between the experimental results and predictions that indicates the detailed consideration of the properties of AHSS that affects the accuracy of the springback prediction.

Most of the studies aim to find the effect of bending parameter to predict the springback for example, Krinninger et al. [18] in their study used different bending parameters to improve the predictability of the springback in a free bending process. The quasi-static parameter variation formed was then used to produce different punch velocities to investigate its influence on the bending process, the springback and the flat length. Erdin and Atmaca [19] used a universal testing machine with 60o

V-bending mold to investigate the effects of holding force on springback behavior of 1050-H14 aluminum alloy plates annealed at 120 C for 20 minutes. The results concluded that annealing decreases springback values in all anisotropy directions and that application of holding force has a significant affirmative effect on springback values. Singh and Agrawal [20] used deformation machining to investigate the comparison of the deforming forces, residual stresses and geometrical discrepancies in Deformation Machining (bending and stretching mode) with conventional and incremental sheet metal bending/stretch forming. The experiment observed that a substantial reduction in deforming forces in deformation machining and incremental bending/forming over conventional bending/forming process.

Despite the many existing study related to springback, there is a lack of study that discuss the effect of applying the bending load at end of loading stage for different time spans and multistage bending (MBT). In this paper, after developing the relationship between the input parameters including modulus of elasticity, sheet thickness and die opening and the output parameter, namely the springback. The paper discuss the effect of the time that the load keep applying at the end of the loading stage.

2. Springback in air bending process

Sheet metal bending process consists of loading and unloading stages as shown in figure 1.

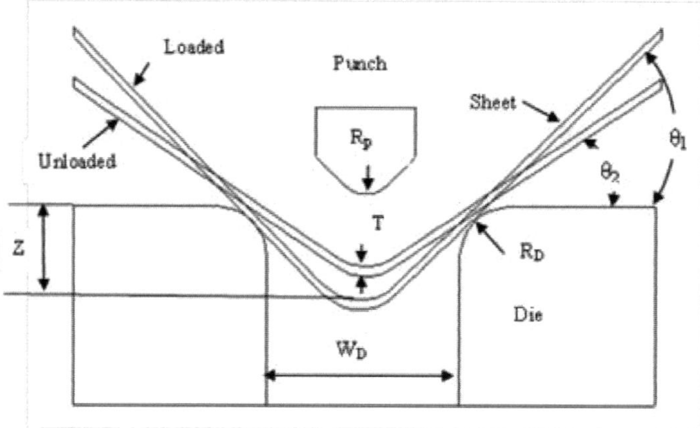

Figure 1. Air bending process.

Due to the mechanical relaxation, the sheet bending angle becomes different after removing the applied load. The change of the angle is called springback and given by equation (1).

$$\Delta\theta = \theta_1 - \theta_2 \tag{1}$$

In this research, the input variables to the ANN metamodel include material parameters and geometrical parameters. The ANN can be defined by the nonlinear relation given in equation (2).

$$\Delta\theta = f(E, \sigma_y, k, n, t, W_d, \theta_1) \tag{2}$$

Where $\boldsymbol{E}$ is the Young' modulus, $\boldsymbol{\sigma_y}$ is the yield strength, $\boldsymbol{k}$ is the strength coefficient, n is the strain hardening coefficient, $\boldsymbol{t}$ is the sheet thickness, $\boldsymbol{W_d}$ is the die opening, and $\boldsymbol{\theta_1}$ is the bend angle prior to springback.

3. Experimental work

3.1. experimental set up

A deformation machine A7 was adapted to apply the air bending process. The die is placed on the machine table, after set up the required sheet, the punch downwards to bend the sheet metal to 90°.

IOP Conf. Series: Materials Science and Engineering **323** (2018) 012021 doi:10.1088/1757-899X/323/1/012021

Stainless steel, Aluminium and brass were selected as target materials for the experiments, table 1 presents the properties of work piece materials.

Table 1. Properties of materials.

Materials	Strength coefficient (MPa) (K)	Strain hardening exponent (n)	Tensile Strength (MPa)	Yield Strength (MPa)
Aluminium	870	0.17	483	345
Brass	585	0.21	400	250
Stainless steel	1400	0.44	515	205

3.2. Design of the Experiments

The experiments were designed to investigate the springback obtained when bending the sheet metals to 90° for different bending parameters. The parameters that were varied in order to assess their effect were: material, sheet thickness and die opening. Each factor had different values, as shown in the table 2.

Table 2. Parameters levels.

		Levels
Modulus of elasticity (GPa)		69, 110, 200
Thickness (mm)	Aluminium	2, 3.5, 6, 8, 10
	Brass	2, 4
	Stainless Steel	2, 4, 6
Die opening (mm)		20, 40, 133

4. Artificial neural network

ANN are computer programs based on the neural structure of the brain. An ANN is based on a collection of connected simple units called artificial neurons which is inspired from the neural structure of the brain. It is used in machine learning and computer science. ANN functions like a human brain, as in it learns from experience. The more experience or data it is provided with the quicker and more extensively it learns hence the more knowledge it gathers; this process is also known as training. It does not learn from programming or codes like usual programs and therefore offers sophisticated, intelligent computations which are similar to the ones that human brain produces. It includes methods, tools and systems devoted to simulating human methods of logical and inductive knowledge acquisition, reasoning of brain activity for solving problems. It helps computers recognize patterns and generalize those patterns of the past into actions of the future. A trained neural network can be thought of as an expert in the category of information it has been given to analyze. This expert can then be used to provide projections given new situations of interest and answer "what if" questions. Neural networks, with their remarkable ability to derive meaning from complicated or imprecise data, can be used to extract patterns and detect trends that are too complex to be noticed by either humans or other computer techniques.

Given the function of neural networks of identifying patterns or trends in data, they are well suited for prediction or forecasting needs, and hence they are successfully used in sales forecasting, data validation, industrial process control, target marketing, customer research, risk management and so on. The specificity of the work it does is in the following paradigms: recognition of speakers in communications; weather prediction; recovery of telecommunications from faulty software; undersea mine detection; texture analysis; three-dimensional object recognition; hand-written word recognition; and facial recognition and so on. Moreover, ANN can also do everything that any computer software can do making it a very useful program that extends around different fields. Like the functions of biological neuron, the artificial neuron is the building component of the ANN designed to simulate similar functions. The artificial neurons are also known as processing elements (PE) as they process information. Each processing element has weighted inputs, transfer function and one output and it is essentially an equation which balances inputs and outputs as shown in figure 2. The arriving signals, called inputs, multiplied by the connection weights are first combined and then passed through a transfer function to produce the output for that neuron [21].

In this study, ANN metamodel was trained using data generated experimentally for different types of materials and geometrical bending tool dimensions. Based on this metamodel, friendly, easy-to-use software can be developed to predict the springback such that the metamodel can be used in production units to reduce the number of trials needed to achieve a required bending angle after springback. The required training data for the metamodel is selected so that it can provide a wide range of information between the inputs and outputs. Twenty two experiments are carried out to generate training examples and the experiments were chosen carefully to avoid inaccurate approximation.

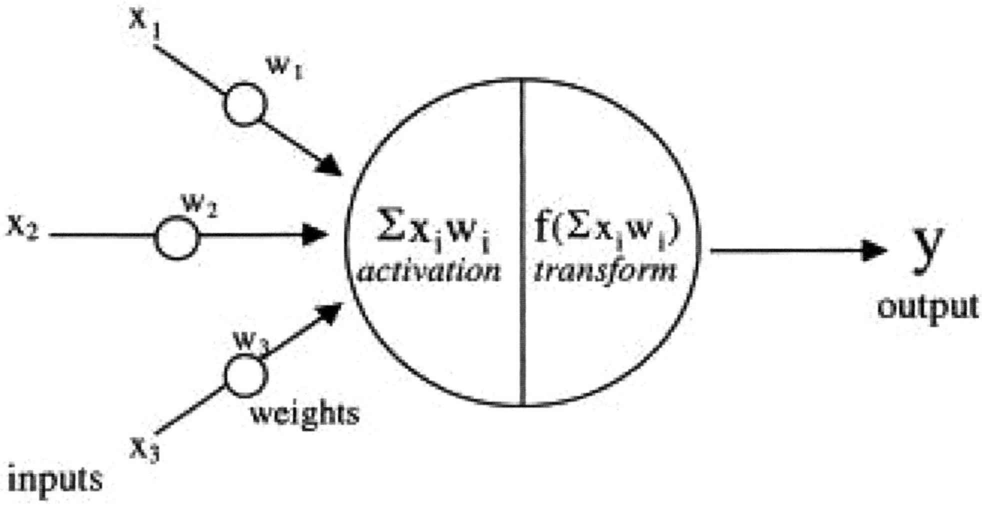

Figure 2. Model of an artificial neuron

These training examples are divided into two groups. The first group consists of 19 training examples and the other group has 3 examples. The second group will be used for validation. To model equation (2) using the ANN metamodel, the ANN architecture and topology such as hidden layers and the number of neurons in each layer should be selected [22]. The number of neurons in the input layer of 3 and one output layer are selected. Back-propagation architecture with one hidden layer is sufficient for the majority of applications [23], therefore, one hidden layer for the neural networks was adopted. In the validation step, the 3 testing examples were used to measure the predictive performance of the metamodels.

5. Results and discussion

5.1. Experimental results

To study the effect of material parameters on the springback amount, one material parameter was varied whereas the other parameters were kept constant. The result is shown in figures 3-5. Figure 3 shows the effect of the modulus of elasticity on the springback amount for a given bending angle before unloading. It is obvious that a lower elastic modulus values produce higher springback because they produce higher strain. Figure 4 shows the effect of the die opening on the springback, it was found that, the springback increases as the die opening increased, also, it was found that, thicker sheet metal leads to lower springback amount as shown in figure 5.

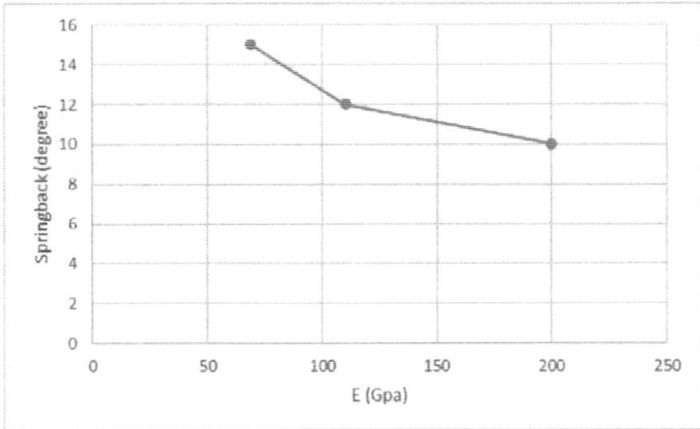

Figure 3. Effect of the modulus on the springback.

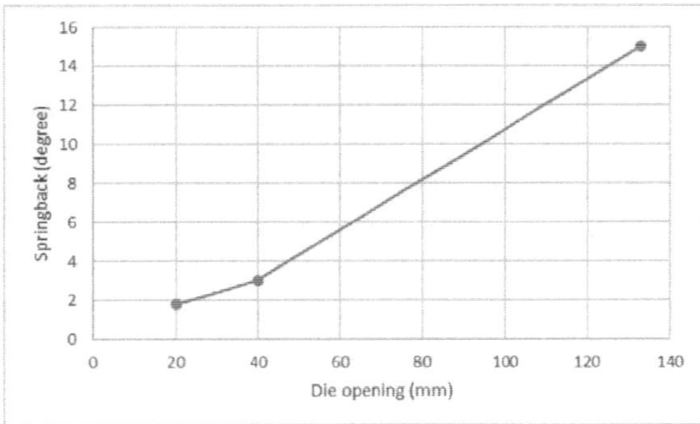

Figure 4. Effect of the die opening on the springback.

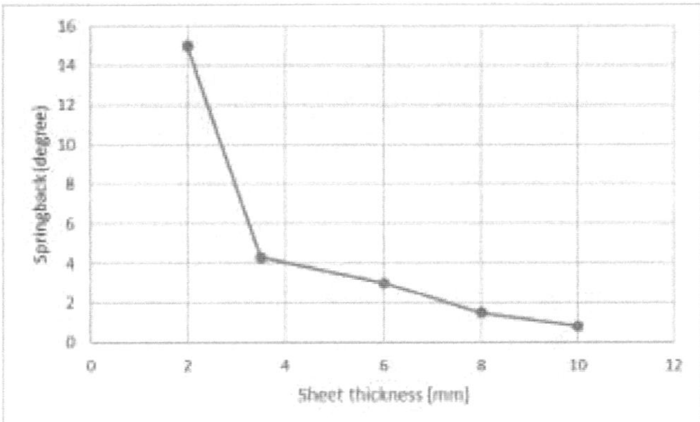

Figure 5. Effect of sheet thickness on the springback.

5.2. ANN results

The ANN results is shown in figure 6. The ANN developed model can be used to predict the springback for unseen input with acceptable accuracy.

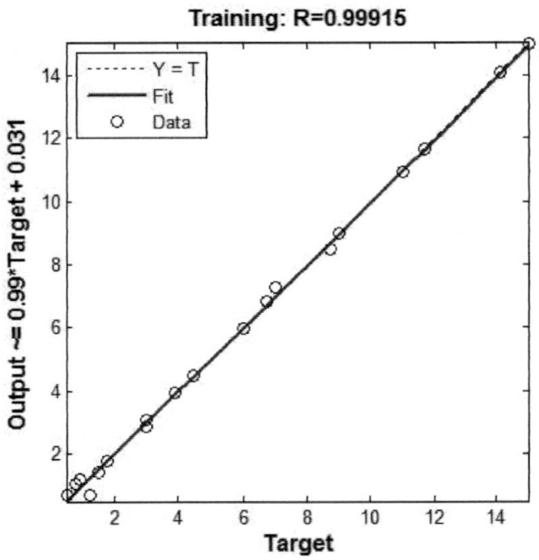

Figure 6. ANN Results.

6. Loading time

The bending process consists of two stage loading and unloading stages. In loading stage, the load is continue applying to achieve the required bending angle. The unloading stage follow the loading stages where the load is released instantaneously after bending the sheet to the required bending angle. This paper investigated the springback under different loading conditions. Instead of release the applied load instantaneously after completing the loading stage, the load was keep applying for a certain period of time. The test was repeating for different period of times and the effect of these time period the springback was investigated. The experiments we carry out with stopping time of 0, 5, 10, 15 minutes. It found that the springback were decreases as the stopping time increased. In stainless steel of 2 mm sheet thickness and under the same parameters it was found that the springback was decreased by approximately 0.5 mm for each stopping time of 5 minutes. Figure 7, explain the direction of the tension and compression stress. In tension side, the space between atoms is larger, so it moves to the more elongated location in the tension side during stopping time.

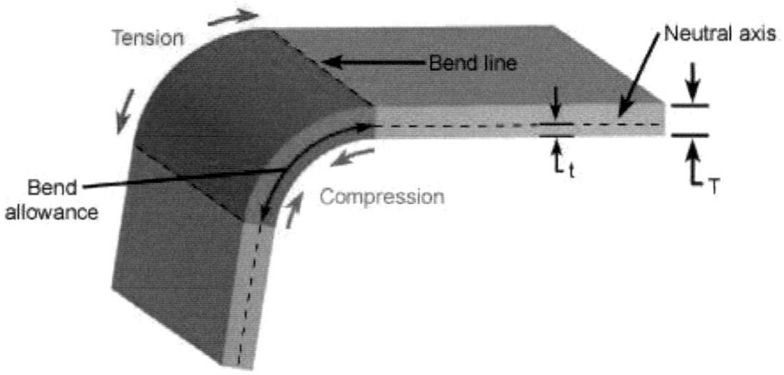

Figure 7. Direction of tension and compression during loading stage

7. Conclusion

The paper investigated a springback in the plan-strain air bending process. To study the effect of material and process parameters on the springback an experimental study were conducted. Based on this study the following remarks are drawn: Springback decreases with the modulus of elasticity because the resistance to elastic bending increases with increasing modulus of elasticity. Springback increases with the yield stress, strength coefficient since the higher these values are the greater the resistance to plastic yielding. Alarge die opening leads to more springback than a smaller one. The most significant parameters effect the springback amount are the sheet thickness, the modulus of elasticity and the die width, respectively. Increase the time at the end of loading stage decreases the springback amount.

Acknowledgments

This project was financially supported by United Arab Emirate University, Summer Undergraduate Research Experience (SURE) PLUS Program

References

[1] Özgü Şenol, Volkan Esat, Haluk Darendeliler, Springback Analysis in Air Bending Process through Experiment based A tificial Neural Networks, Procedia Engineering, Volume **81**, Pages 999-1004, 2014.

[2] Z.Q.Jiang, H.Yang, .M. Zhan, X. D. Xu,.G. J. Li, Coupling effects of material properties and the bending angle on the springback angle of a titanium alloy tube during numerically controlled bending, Materials & Design, Volume **31**, Issue 4, Pages 2001-10 , 2010.

[3] Han XIAO, Shi-hong ZHANG, Rong ZHOU, De-hong LU, Springback characteristics of AZ31 magnesium alloy as-extruded profile in warm tension-rotation bending process, Transactions of Nonferrous Metals Society of China, Volume **22**, Supplement 2, Pages s416-s421, 2012.
[4] Heng LI, Kai-peng SHI, He YANG, Yu-li TIAN, Springback law of thin-walled 6061-T4 Al-alloy tube upon bending, Transactions of Nonferrous Metals Society of China, Volume 22, Supplement 2, Pages s357-s363, 2012.
[5] Lifei Wang, Guangsheng Huang, Hua Zhang, Yanxia Wang, Liang Yin, Evolution of springback and neutral layer of AZ31B magnesium alloy V-bending under warm forming conditions, Journal of Materials Processing Technology, Volume **213**, Issue 6, Pages 844-50, June 2013.
[6] Komgrit Lawanwomg, Hiroshi Hamasaki, Ryutaro Hino, Fusahito Yoshida, A Novel Technology to Eliminate U-bending Springback of High Strength Steel Sheet by Using Additional Bending with Counter Punch, Procedia Engineering, Volume **81**, Pages 957-62, 2014.
[7] Wiriyakorn Phanitwong, Sutasn Thipprakmas, Development of anew spring-back factor for a wiping die bending process, Materials & Design, Volume **89**, 5 Pages 749-58, 2016.
[8] Gisario, M. Barletta, S. Venettacci, Improvements in springback control by external force laser-assisted sheet bending of titanium and aluminum alloys, Optics & Laser Technology, Volume **86**, December 2016, Pages 46-53
[9] Ossama Mamdouh Badr, Bernard Rolfe, Peng Zhang, Matthias Weiss, Applying a new constitutive model to analyse the springback behaviour of titanium in bending and roll forming, International Journal of Mechanical Sciences, Volumes **128–129**, Pages 389-400, 2017.
[10] Mehmet Alper Sofuoğlu, Selim Gürgen, Fatih Hayati Çakır, Sezan Orak, Springback Behavior of AA6082T6 Tubes in Three-point Bending Operation, Procedia Engineering, Volume **182**, Pages 658-64, 2017.
[11] Annamaria Gisario, Mehrshad Mehrpouya, Simone Venettacci, Massimiliano Barletta, Laser-assisted bending of Titanium Grade-2 sheets: Experimental analysis and numerical simulation, Optics and Lasers in Engineering, Volume **92**, Pages 110-19, 2017.
[12] S Agatonovic-Kustrin, R Beresford, Basic concepts of artificial neural network (ANN) modeling and its application in pharmaceutical research, Journal of Pharmaceutical and Biomedical Analysis, Volume **22**, Issue 5, 2000, Pages 717-27
[13] M.V Inamdar, P.P Date, U.B Desai, Studies on the prediction of springback in air vee bending of metallic sheets using an artificial neural network, Journal of Materials Processing Technology, Volume **108**, Issue 1, Pages 45-54, 2000.
[14] Feifei Song, He Yang, Heng Li, Mei Zhan, Guangjun Li, Springback prediction of thick-walled high-strength titanium tube bending, Chinese Journal of Aeronautics, Volume **26**, Issue 5, October 2013, Pages 1336-45
[15] Y. Song, Z. Yu, Springback prediction in T-section beam bending process using neural networks and finite element method, Archives of Civil and Mechanical Engineering, Volume **13**, Issue 2, June 2013, Pages 229-41.
[16] Maati, L. Tabourot, P. Balland, E.H. Ouakdi, M. Vautrot, N. Ksiksi, Constitutive modelling effect on the numerical prediction of springback due to a stretch-bending test applied on titanium T40 alloy, Archives of Civil and Mechanical Engineering, Volume **15**, Issue 4, September 2015, Pages 836-46
[17] Xi Yang, Changhyok Choi, Nimet Kardes Sever, Taylan Altan, Prediction of springback in air-bending of Advanced High Strength steel (DP780) considering Young's modulus variation and with a piecewise hardening function, International Journal of Mechanical Sciences, Volume **105**, January 2016, Pages 266-72.
[18] Michael Krinninger, Daniel Opritescu, Roland Golle, Wolfram Volk, Experimental Investigation of the Influence of Punch Velocity on the Springback Behavior and the Flat Length in Free Bending, Procedia CIRP, Volume **41**, Pages 1066-71, 2016.
[19] Muhammed Emin Erdin, Alper Atmaca, Effects of Holding Force on the Springback Behavior of Annealed Aluminum Plates, Procedia Engineering, Volume **149**, Pages 56-61, 2016.

[20] Arshpreet Singh, Anupam Agrawal, Comparison of deforming forces, residual stresses and geometrical accuracy of deformation machining with conventional bending and forming, Journal of Materials Processing Technology, Volume **234**, August 2016, Pages 259-71.
[21] Kleijnen, J.P.C. and R.G. Sargent, A methodology for fitting and validating metamodels in simulation. European Journal of Operational Research European Journal of Operational Research120(1): Pages 14-29, 2000.
[22] Haykin, S.S., Neural networks and learning machines, New York: Prentice Hall/Pearson, 2009.
[23] Hornik, K., M. Stinchcombe, and H. White, Multilayer feedforward networks are universal approximators. Neural Networks. 2(5): Pages 359-366, 1989.

IOP Conf. Series: Materials Science and Engineering **323** (2018) 012022 doi:10.1088/1757-899X/323/1/012022

The charging security study of electric vehicle charging spot based on automatic testing platform

Yulan Li[1*],Zhangli Yang[2] Bin Zhu[1] and Shengyi Ran[1]

[1]Chongqing College of Electronic Engineering, No. 76, University Town East Road, Shapingba District, Chongqing, China
[2]State Grid Chongqing Elect Power Co, Elect Power Res Inst, Chongqing, China

*Corresponding author: 153033107@qq.com

Abstract. With the increasing of charging spots, the testing of charging security and interoperability becomes more and more urgent and important. In this paper, an interface simulator for ac charging test is designed, the automatic testing platform for electric vehicle charging spots is set up and used to test and analyze the abnormal state during the charging process. On the platform, the charging security and interoperability of ac charging spots and IC-CPD can be checked efficiently, the test report can be generated automatically with No artificial reading error. From the test results, the main reason why the charging spot is not qualified is that the power supply cannot be cut off in the prescribed time when the charging anomaly occurs.

1. Introduction

According to the guidelines for the development of electric vehicle charging infrastructure (2015-2020 years) ([2015]1454), by 2020, the number of replacement power plants will reach 12 thousand, and charging spots will reach 4 million 500 thousand. Also in China, electric vehicles and their infrastructure showed explosive growth, the industry has entered the fast lane.

However, the interconnection and product security problems that have been plaguing the industry are becoming more and more prominent. It is necessary to carry out systematic test on charging security and interoperability of electric vehicle charging spots [1]. Before 2015, the detection of charging spots was generally adopted by industry standard NB/T 33008-2013 [2-3]. At the end of 2015, 5 national standards for electric vehicle charging interface and communication protocol was promulgated by AQSIQ and the National Standard Committee jointly [4-6], the compatibility provisions was further refined, the communication version control was added, the charging timing logic and time definition were defined and so on. The latest test specifications for conduction charging interoperability testing rules, condition, items and qualifying the judgment has been carried on [7].

At present, there are few studies on the testing of charging facilities, and there are few relevant research literatures. In literature [1], the necessity of conformance testing and interoperability testing of electric vehicle charging facilities is analyzed. The schematic diagram of the test structure and the application in the power industry, especially the smart grid are introduced. The paper introduces the research and design of a mobile testing platform for the electric vehicle charging facilities [8], carries out the demand analysis, the scheme design, draws the test diagram. In literature [9], the basic functions and working principle of the charging control circuit of electric vehicle are introduced. A method of matching resistance and voltage state test in different charging modes is designed. In a

IOP Conf. Series: Materials Science and Engineering **323** (2018) 012022 doi:10.1088/1757-899X/323/1/012022

word, nowadays the charging facilities electrical performance and interoperability testing mainly rely on testing staff manual one by one to detect, manually read the instrument data, and then make a judgment [10]. Therefore, for detection efficiency on the electric vehicle charging facilities is low, heavy workload, prone to human error, testing personnel demanding high, the automatic detection platform of electric vehicle charging spots is designed and built, and the abnormal states that may occur during the charging process are tested and analyzed according to the new national standard and the latest interoperability test specification.

2. Hardware platform

The design schematic diagram of the automatic test platform for electrical car charging spots is shown in figure 1 below. On the platform, General electric characteristics and interoperability of ac charging spot can be tested automatically. The platform contains these following equipment, such as Programmable ac power supply, Programmable AC load, Ac charging test interface simulator, The oscilloscope, High precision power meter, High performance data acquisition card, Serial server, Signal selector, Industrial computer and so on. Automatic detection platform is shown in figure 2.

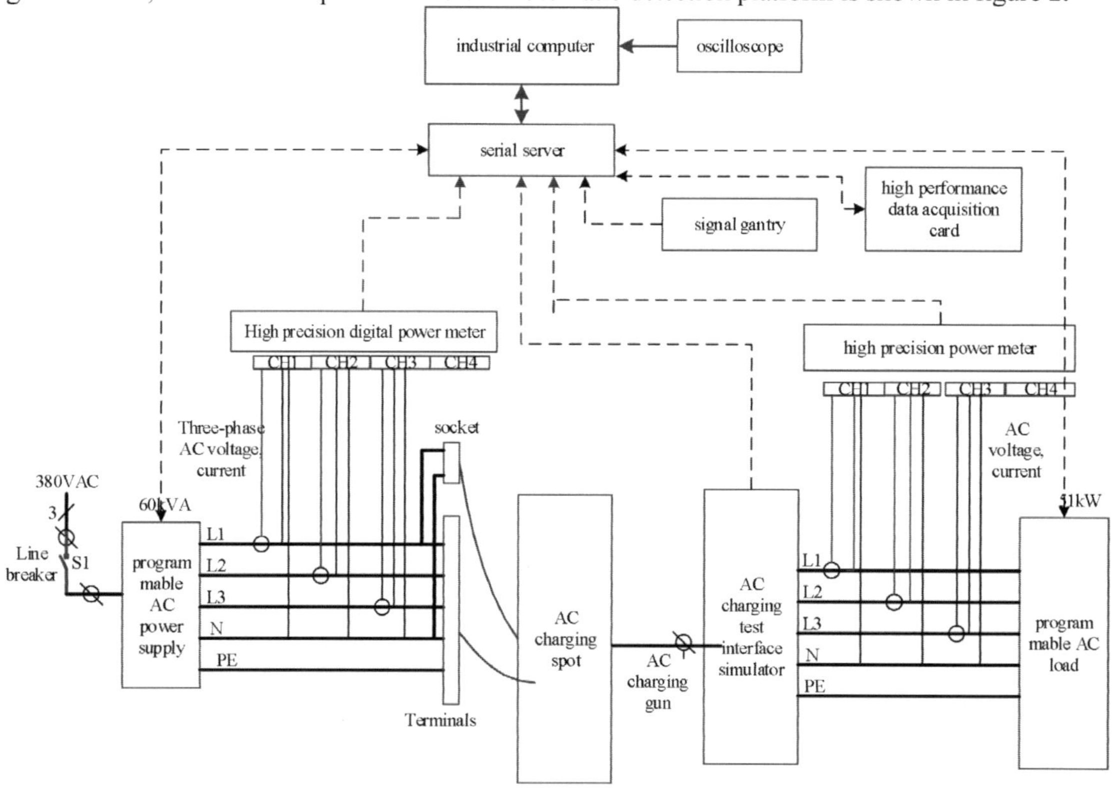

Figure 1. Test equipment schematic diagram of A.C. charging spot and in-cable control and protection device (IC-CPD).

Figure 2. Automatic detection platform.

3. Ac charging test interface simulator

Schematic diagram of the ac charging test interface simulator is shown in figure 3.On the device, the changes in guidance resistor, the faults of signal grounded or broken and switch on-off logic can be simulated, Electrical characteristics and control guidance tests of AC piles can be finished under the coordination of power supply and load, even the compatibility and correct exit for protection action of the electric vehicle can be tested . The device is also provided with a dedicated socket and an interface of IC-CPD and the AC pile input power, so the simulation test of the IC-CPD and the AC pile can be finished.

Simulation methods for various situations and detailed implementation are as follows,

The insulated analog resistor and the analog switch are connected between the L1 and the PE, if the insulation resistance reduction occurs, the leakage protection action of the charging spot is caused, and the residual current function can be tested.

The ground resistance and the analog switch are connected between the CC and the PE to simulate CC grounding fault, if they are connected between the CP and the PE, CE grounding fault is simulated.

The R2 resistor analog circuit can simulate the 5 resistor positions of the R2 resistor, 909, 1261, 1300, 1339 and 1723 ohms respectively.

The R3 resistor analog circuit can simulate the 5 resistor positions of the R3 resistor, 1820, 2658, 2740, 2822 and 4610 ohms respectively, resistance accuracy 0.2%F.S., power >0.5W.

The vehicle control simulator is composed of DSP and peripheral circuits, DSP is the core component, and the peripheral circuits is composed by the IO input and output and analog acquisition. The vehicle control simulator can also communicate with secondary centralized control unit by 485 interface and Receive instructions and return the internal state. Each contact including L1, L2, L3, N, PE, CP and CC is equipped with analog switches, which can simulate the fault state.

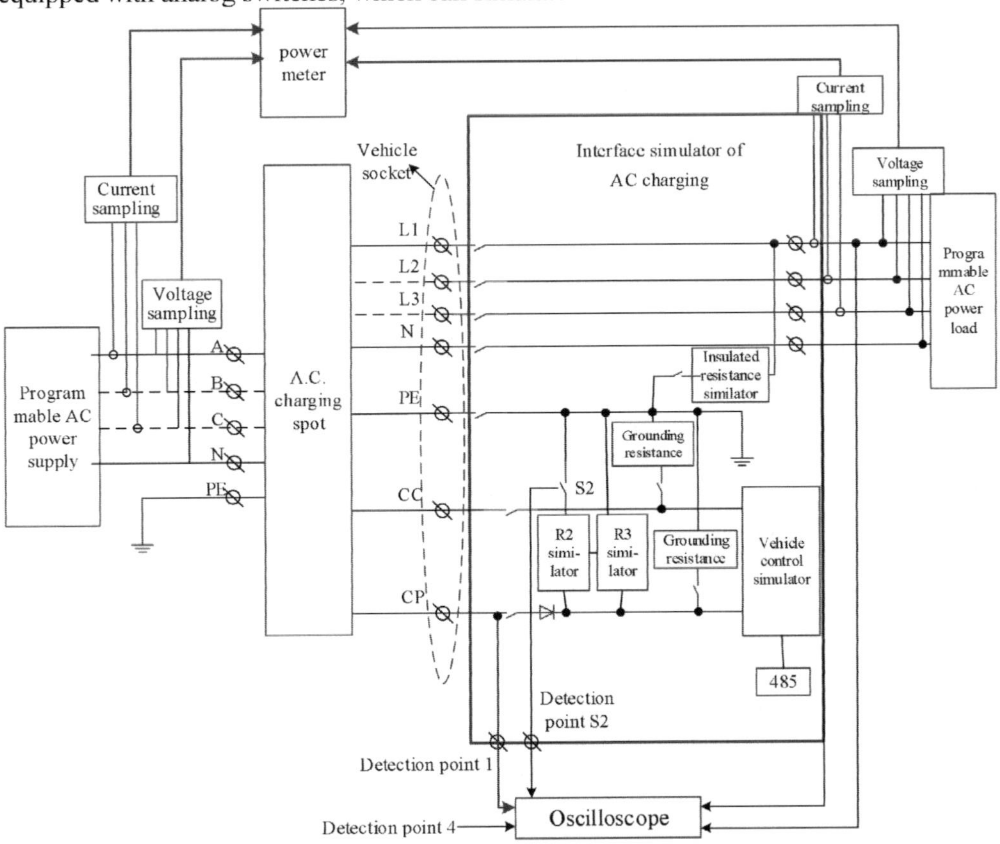

Figure 3. Test equipment schematic diagram of A.C. charging spot and in-cable control and protection device (IC-CPD).

IOP Conf. Series: Materials Science and Engineering **323** (2018) 012022 doi:10.1088/1757-899X/323/1/012022

4. Test items and results

According to the test specification of electric vehicle conduction charging interoperability, part 1: power supply equipment (press release), The interoperability test items and IC-CPD for ac charging spots are shown in table 1. This paper tests the abnormal conditions during the charging process, and the test results are described below.

Table 1. Test items be carried out.

Number	Test items	remarks
A1.1001/I1.1001	Connection confirmation test	The test point 4 only for connection mode B
A1.3001/I1.3001	Charge preparation test	
A1.4001/I1.4001	Start-up and charging stage test	
A1.5001/I1.5001	Normal charging end test	
A1.6001/I1.6001	Charging connection control timing test	
A1.3501/I1.3501	CC disconnection test	only for connection modeA or B
A1.3502/I1.3502	CP disconnection test	
A1.4501/I1.4501	CP Grounding test	
A1.4502/I1.4502	Protective grounding conductor continuity loss test	
A1.4503/I1.4503	Output overcurrent test	
A1.6002/I1.6002	CP loop voltage limit test	

4.1. CP disconnection test

The item checks whether the charging pile can stop charging when the CP is disconnected, before charging and charging, respectively. According to the standard, if the CP disconnection occurs in charging, the power supply control device should disconnect ac power supply circuit within 100 ms, if the CP disconnection occurs before charging, The charging control switch S1 switch to +12V ,and does not close the power supply loop. The CP broken test results are shown in table 2.

Table 2. Test results of simulation CP disconnection.

	Test point 1			Stop charging time (S)
	High level (V)	Low level(V)	Frequency(Hz)	
Before charging	12.960	11.040	6779.700	-
charging	12.960	10.800	5970.100	0.069

4.2. CP grounding test

The item checks whether the charging pile can stop charging when the CP is grounded, before charging and charging, respectively. According to the standard, if the CP grounding occurs in charging, the power supply control device should disconnect ac power supply circuit within 100 ms. The results of simulation CP grounding are shown in table 3. it can be seen that the Stop charging time is out.

Table 3. Test results of simulation CP grounding.

Test point 1			Stop charging time (S)
High level (V)	Low level(V)	Frequency(Hz)	
2.400	0.720	2020.200	0.585

4.3. Protective grounding conductor continuity disconnection test

Whether the charging spot can stop charging can be checked in this item when the Protective grounding conductor continuity loss. Normally, Electric vehicle power supply equipment should disconnect power supply within 100 ms. The results are shown in table 4, and meet the standard requirements.

Table 4. Test results of protective grounding conductor continuity disconnection.

	Test point 1			Stop charging time (S)
	High level (V)	Low level(V)	Frequency(Hz)	
Before charging	12.960	11.040	1960.800	-
charging	12.960	10.800	1754.400	0.068

4.4. Output overcurrent test

Whether the charging spot can stop charging can be checked when output overcurrent occurs. When the PWM signal is generated from the charging spot, the corresponding maximum power supply current is less than 20A and the amplitude of simulated charging current is 2A more than the maximum power supply current of the charging spot and maintains 5s, or the corresponding maximum power supply current is more than 20A and the amplitude of simulated charging current is 1.1 times higher than the maximum power supply current of the charging spot and maintains 5s, the output supply power should be disconnected and the switch S1 should be switched to the +12V connection state .From table 5, The test sample is still charging when at 9.636S and does not meet the requirements.

Table 5. Test results of output overcurrent.

Test point 1			Output voltage (V)	Output current (A)	Stop charging time (S)
High level (V)	Low level(V)	Frequency(Hz)			
5.760	-12.240	1000.000	212.920	36.500	9.636

4.5. Switch disconnection state test

Whether the charging spot can stop charging can be checked when the switch s2 is switched off here. When the voltage amplitude of test point is 9V, S2 is switched off, the power supply control device should disconnect ac power supply circuit within 100ms according to the requirement and generate PWM output continuously. From table 6, the sample meets the requirements.

Table 6. Test results of disconnect switch S2.

Test point 1			Stop charging time (S)
High level (V)	Low level(V)	Frequency(Hz)	
8.880	-12.240	1000.000	0.055

4.6. Loop voltage limit test

The voltage value of the test point 1 is normal or not represents the charging spot is charging or not. If the positive voltage value at the detection point 1 is within the range of nominal value of the corresponding state, the charging spot should be allowed to charge or recharge normally, otherwise the charging spot should not be allowed to charge or recharge. From table 7, the sample meets the requirements.

Table 7. Test results of loop voltage limit.

	Test point 1			Output voltage (V)	Output current (A)	Stop charging time (S)	remarks
	High level (V)	Low level(V)	Frequency(Hz)				
Before charging	10.800	8.880	5263.200	-	-	-	Upper limit
charging	6.960	-12.240	999.938	220.360	0.000	0.544	
Before charging	8.640	6.960	651.466	-	-	-	Lower limit
charging	5.760	3.840	490.196	0.000	0.000	0.584	

5. Conclusion

Ac charging test interface simulator is designed and automatic test platform for charging spots of electric vehicle is set up in the paper. This platform is used to test and analyze the abnormal state which may occur during the charging process. On the platform, the charging security and interoperability of ac charging spots and IC-CPD can be checked efficiently, the test report can be generated automatically with No artificial reading error. From the test results， the main reason why the charging spot is not qualified is that the power supply cannot be cut off in the prescribed time when the charging anomaly occurs.

Reference

[1] LIU Hai-bo1, LI Jiang-xiang1, YUAN Hong, HAN Yuan-kai, etl. Information interoperability technology research for electric vehicle charging/battery swap infrastructure. *Chinese Journal of Power Sources*, vol. **12**, pp. 2449-52, 2016.

[2] NB/T 33008.1-2013, *Inspection and testing specifications for electric vehicle Charging equipment Part 1: off-board charger.*

[3] NB/T 33008.2-2013, *Inspection and testing specifications for electric vehicle Charging equipment Part 2: AC charging spot.*

[4] GB/T 18487.1-2015, *Electric vehicle conductive charging system Part 1: General requirements* [S].

[5] GB/T 20234.2-2015, *Connection set for conductive charging of electric Vehicles Part2: AC charging coupler.*

[6] GB/T 20234.2-2015, *Connection set for conductive charging of electric Vehicles Part2: AC charging coupler.*

[7] GB/T XXXXX-2016, *Interoperability test specifications of electric vehicle conductive charging Part 1: Supply equipment* .

[8] HU Chun-yu, LU Yi, LI Xiang, HUANG Xiao-ming, LI Wu-feng. *Research and design of charging spots mobile detection platform. Chinese Journal of Power Sources*, vol. **03**, pp. 697-699, 2016.

[9] SANG Li, XU Hong-Hai, GUAN Xiang. Experiment Design on Control Pilot Circuit of AC Charging Coupler for Electric Vehicle . *Electrical Measurement & Instrumentation*, vol. **02**, pp. 112-115+120, 2013.

[10] DING Xin-zhi, BI Zhi-zhou, CAO Min, et al. Technique Analysis and Experiment of Electric Vehicle Off-board Charger. *Electrical Measurement & Instrumentation*, vol. **04**, pp. 14-17, 2012.

IOP Conf. Series: Materials Science and Engineering **323** (2018) 012023 doi:10.1088/1757-899X/323/1/012023

Two layers LSTM with attention for multi-choice question answering in exams

Yongbin Li[1,2]

[1]ZunYi Medical University, No.6 Xuefu West Road, Xinpu new distract, Zunyi, GuiZhou, CHINA
[2]YunNan University, University City, Chenggong distract, Kunming, YunNan, CHINA

*Corresponding author's e-mail: 1957405@qq.com

Abstract. Question Answering in Exams is typical question answering task that aims to test how accurately the model could answer the questions in exams. In this paper, we use general deep learning model to solve the multi-choice question answering task. Our approach is to build distributed word embedding of question and answers instead of manually extracting features or linguistic tools, meanwhile, for improving the accuracy, the external corpus is introduced. The framework uses a two layers LSTM with attention which get a significant result. By contrast, we introduce the simple long short-term memory (QA-LSTM) model and QA-LSTM-CNN model and QA-LSTM with attention model as the reference. Experiment demonstrate superior performance of two layers LSTM with attention compared to other models in question answering task.

1. Introduction

Question Answering in Exams is typical question answering task can be formulated as follows: Given a question q and an answer candidate pool { a1 , a2 , ⋯ , as } for the question (s is the pool size, in this task is 4), we aim to find the best answer candidate ak , $1 \leq k \leq s$. Question and answer are token sequence with arbitrary length, and a question can correspond to multiple ground-truth answers, and correct answer may only have semantically relation with question instead of directly sharing lexical units. The selected answer ak is inside the ground truth set or outside, from the definition, the Question Answering task can be transform as a binary classification problem. For each question, for each answer candidate, we construct a QA pair, it may be appropriate or not, we need to measure the matching degree, the highest will be chosen.

The above is general definition. The task is come from a NLP competition, we employ multiple choice questions from a typical science and history curriculum, questions are restrained within the elementary and middle school level. For each question, there are four possible answers, where each of them may be a word, a value, a phrase or even a sentence.

We approach is constructing a question answers pairs for each question and his one answers, in train dataset, we need to mark this is a positive sample or negative sample. In validation and test dataset, we employ the same processing mode, determining the matching degree of fit between a question and an answer, the highest will be chosen. The deep learning (DL) models are based on utilizing word distributed representation on both questions and answers, and building two layers long short-term memory (LSTM) respectively, on top of this, we introduce an efficient attention model for the answer embedding generation according to the question context, finally, a dense layer is used to determine the matching degree. For confirming the effect of the model, we also tested the QA-LSTM-CNN model and QA-LSTM with attention model to demonstrate the validity of two layers LSTM with attention. By experiment, we use two layers LSTM with attention model get a better performance.

2. Related work

Previously, feature extraction, semantic analysis was usually used in question answering task. For example word semantic relations were makeup based on WordNet in [1]. In [2] [3] using syntactical matching. [4] tried to fulfill the matching using minimal edit sequences between dependency parse trees. Above methods show effectiveness, but these relies on feature extraction and language tools.

With the development of neural networks, we adopt the method of word distributed representation from [5], LSTM model from [6], and the CNN model refers to paper [7]. State-of-the-art, there are three main directions to solve Question Answering Problems based on deep learning models: Firstly, the question and answer representations are learned and matched by similarity metrics [8] [9]. Secondly, a joint feature vector is constructed based on both the question and the answer, then task was converted into a classification problem [10]. Finally, the proposed models for textual generation can intrinsically be used for question answering task and generation [11] [12].

The approach in this task is based on the second category, attempting to transform this task into a classification task, using LSTM models and attention models. The model framework is a little similar to [9], but has distinct differences: first, our models employ a method by constructing joint feature vector sending to a dense layer, meanwhile, In order to extract features better, we have adopted a two-layer LSTM model with attention model, and achieved better performance.

3. Model description

In the section, we describe the two layers LSTM with attention model address the task. The following, introduce baseline models which use QA-LSTM models on both questions and their answer candidates and QA-LSTM-CNN and QA-LSTM-attention as the reference models. The structure of two layers LSTM with attention framework are shown in figure 1.

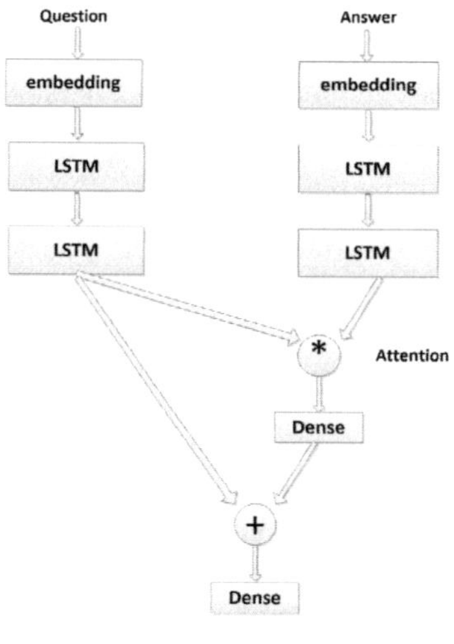

Figure 1. The structure of two layers LSTM with attention.

We train a word embedding by word2vec [2]. The word embedding provides the distributed representation for each token in question and answer respectively. The vectors have dimensionality of 300 and were trained using continuous bag-of-words architecture. Word embedding are parameters that can also be optimized during training. In preliminary experiment, there are two word distributed representation models we used. One is the word embedding from the pre-trained Word2Vec model which were trained by [12] on 100 billion words of Google News and are publicly available, word vector was initialized from an unsupervised neural language model. The other word2vec model is trained by supervised neural language model by a scientific knowledge story corpus which have about 600 thousand story. The format of each story is question and correct answer pairs, e.g. (question, correct answer), the parameter min count is 1, and the window is 10, words not present in the set of pre-trained words are initialized zeros. Through experiments, it is found that the word vector model trained by special scientific knowledge story corpus is more accurate, so the model is chosen in the end.

Each question and answer will be transformed to a word vector matrix and be entered into two layers LSTM models respectively. LSTM is a special type of RNN that can learn to rely on long-distance history and the immediate previous hidden vector, it's a remarkable variations of RNN to alleviate the gradient vanish problem. LSTM was proposed by [6] and has been improved and promoted by [13]. Given an input sentence sequence $x = \{ x_{(1)}, x_{(2)}, \cdots, x_{(n)} \}$, $x_{(t)}$ is the E dimension word vector in t time. The hidden vector $h_{(t)}$ (size is H) at the time step t is updated as follows.

$$i_t = \sigma(W_i [h_{(t-1)}, x_{(t)}] + b_i) \quad (1)$$

$$f_t = \sigma(W_f [h_{(t-1)}, x_{(t)}] + b_f) \quad (2)$$

$$o_t = \sigma(W_o [h_{(t-1)}, x_{(t)}] + b_o) \quad (3)$$

$$\widetilde{C}_t = \tanh(W_c [h_{(t-1)}, x_{(t)}] + b_c) \quad (4)$$

$$C_t = i_t * \widetilde{C}_t + f_t * C_{t-1} \quad (5)$$

$$h_t = o_t * \tanh(C_t) \quad (6)$$

In the LSTM architecture, there are three gates (input i, forget f and output o), and a cell memory vector C. σ is sigmoid function. The input gate can determine how incoming vectors $x_{(t)}$ later the state of the memory cell. The output gate can allow the memory cell to have an effect on the outputs. Finally, the forget gate allows the cell to remember or forget its previous state. $[h_{(t-1)}, x_{(t)}]$ mean to concatenate vector $h_{(t-1)}$ and vector $x_{(t)}$, $W \in R^{H\times(H+E)}$, $b \in R^H$.

Questions and answers were respectively processed through the embedded layer and the two layers LSTM model, generate two sequences of LSTM output vectors. Now, we investigate a state-of-the-art attention model for the answer vector generation based on question, instead of generating QA representation respectively. If the input sentence is longer, semantics are expressed by an intermediate semantic vector, and the information of the word itself has disappeared, which results in the loss of a great deal of detail information. An attention mechanism are used to alleviate weakness by dynamically aligning the more informative parts of answer to the questions. This strategy are widely employed in many natural language processing tasks, such as machine translation and so on. We develop the efficient word-level attention above the two layers LSTM model. Answer output vector will be multiplied by a softmax weight, which is determined by the question outputting from LSTM. Specifically, attention model give more weights on certain words, just like tf-idf for each word, however, the weight is calculated by the question. Given the output vector of two layers LSTM on the answer side at time step t, $h_a(t)$, and the question embedding, o_q, the updated vector $h_a(t)$ for each answer token are formulated below.

$$m_{a,q}(t) = \tanh(W_{am} h_a(t) + W_{qm} o_q) \quad (7)$$

$$s_{a,q}(t) \propto \exp(W_{m,s}^T m_{a,q}(t)) \quad (8)$$

$$\tilde{h}_a(t) = ha(t) s_{a,q}(t) \quad (9)$$

Final, we merge the two sequences to one and perform flatten operations, the ultima output is passed through a two-dimensional softmax layer.

As reference, we apply a baseline QA-LSTM model and QA-LSTM-CNN model and QA-LSTM-attention. In QA-LSTM-CNN, we resort a CNN and MaxPooling built on every outputs of LSTM. We find that there is a certain improvement in accuracy compared with basic QA-LSTM model, it is because model can obtain a more composite n-gram representation of questions and answers. Convolutional structure only imposes local interactions between the inputs within a filter size m, size 3 filters are used here. Every windows with size of m order to capture m line information in LSTM output matrix, after the convolution operation, we have added a pooling layer which adopt MaxPooling, pool size is 2. Intuitively, the pooling operation can reduce the amount of computation of the model while retaining most of the information. In preliminary experiment, we do not see a significant difference in accuracy between different pooling methods, so MaxPooling is used only. The QA-LSTM-attention employ a single layer of LSTM with attention which achieves the same accuracy as the QA-LSTM-CNN model, different LSTM units are used here.

We also tried to insert CNN layers before the LSTM with attention framework and found the results were not good, and finally observe it better to replace the CNN with LSTM. Our guess is that the CNN network needs to pad a large number of zeros, resulting in a large amount of invalid data is created, reduces the capability of semantic feature extraction.

4. Related dataset

The task is come from a NLP competition, the train dataset consists of two parts: science and history curriculum dataset, SciQA and 8th Studystack flashcards.

Science and history curriculum: The dataset is a collection of American middle school science and history curriculums selected by competition organizers. A total of seven subsets (e.g. biology, chemistry) are included in this subjects, the trainset contains 3886 questions, and each questions have 4 answer candidates.

SciQA and 8th Studystack flashcards: SciQA and 8th Studystack flashcards is an additional train dataset corpus we introduced, download from https://www.studystack.com/. The dataset is also a multiple-choice question for science and history courses from various regional and national science examinations, there is 13455 sample which have the same for format as above

The validation dataset and test dataset are also scientific and historical questions, with the quantity being 669 and 812 respectively, these question not duplicate from the train dataset.

5. Experimental setup

The models in this task are implemented with keras, we use the accuracy on validation dataset to locate the best epoch and best hyper-parameters for test. The final rate of accuracy is expressed in the correct proportion chosen, formula is as follows:

$$accuracy = \frac{number\ of\ correct\ questions}{total\ number\ of\ questions} \tag{10}$$

We apply word2vec model which is trained by a scientific knowledge story corpus which have about 600 thousand story instead of the pre-trained google news Word2Vec model, and the vector dimensionality is 300. Word vectors are also parameters and can be optimized during training. We use the loss function of categorical cross entropy and the optimizer of adaptive moment estimation. We tried to include l_2 norm in preliminary experiments, but the regularization factors doesn't show any improvement, so we just used dropout layers to prevent overfitting, the parameter is 0.3.

The max length of questions and answers tokens sequence of training datasets is limited to 64, any tokens out of this range will be discarded. If the length of tokens sequence is not enough, then zero is added.

For comparison, we report the performance and analysis of four framework in table 1. The table summarizes the results of our models for the question answering task. From Row (1) to (2), we list the baseline QA-LSTM without either CNN structure or attention mechanism. We use epochs for 20, batch size for 512, and merge mode for sum, the parameter mask_zero means masking for input data with a variable number of time steps. The difference is that the different output dimensionality parameters of LSTM for questions or answers are used, we can observe that increasing the number of units helps to improve accuracy. We take the results of the simple LSTM model as baseline to measure other models.

From Row (3) to (4), CNN and MaxPooling layers are built on the top of the LSTM with different filter numbers with 100 or 200, we set the filter size on 3, we did not see better result if we increase m. The size of pooling operation is 2, and merge mode is concatenate. From the results of the LSTM-CNN model, we can perceive that the addition of CNN structure is beneficial to improve the capability of the model.

Row (5) correspond to QA-LSTM with the attention mechanism, output dimensionality parameters we are still using 64. We observe that the improvement form attention is remarkable, we get improvements from the baseline of simple LSTM model by 6.5% and 6.9% in validation and test dataset, compared to the QA-LSTM-CNN model, it also improved by more than 2%.

Row (6) is the framework proposed in this paper, two layers LSTM with attention. As can be seen from the result, the models get a significant result, achieve a precision of 37.67%. Compared to the single layer LSTM-attention model, it also improved by more than 2%.

Table 1. The results of four models on the question answering task.

Idx	Model	validation	Test
1	QA-LSTM baseline(output_dim=32, mask_zero=True)	28.12	29.14
2	QA-LSTM baseline(output_dim=64, mask_zero=True)	29.99	31.01
3	QA-LSTM-CNN(output_dim=64, filters=100, kernel_size=3 pool_size=2, border_mode="valid")	30.31	32.56
4	QA-LSTM-CNN(output_dim=64, filters=200, kernel_size=3 pool_size=2, border_mode="valid")	31.99	33.22
5	QA-LSTM-attention(output_dim=64)	34.61	36.02
6	two layers LSTM with attention(output_dim=64)	37.11	37.67

6. Conclusion

In this paper, we study the answer selection task by employing a two layers LSTM with attention models based deep learning framework. The proposed approach is to build distributed word embedding of question and answers which is trained by a scientific knowledge story corpus instead of the pre-trained google news Word2Vec model, meanwhile, does not rely on manually extracting features or linguistic tools, and can be applied to many domain. We further introduce attention mechanisms to solve question answering task, the answer vector generation based on question, instead of generating QA representation respectively, the weighted transformation can effectively improve the performance of sequences under Natural Language Processing. On this basis, we find that two layers LSTM with attention can achieve better performance compared to other framework. We conduct experimental using science and history curriculum dataset, SciQA and 8th Studystack flashcards, our experimental results demonstrate that the proposed model outperform a variety of other framework. Therefore, the two layers LSTM with attention model is the framework recommended by this paper for question answering task.

References

[1] Yih, Wen-tau, Chang, Ming-Wei, Meek, Christopher, and Pastusiak, Andrzej. Question answering using enhanced lexical semantic models. Proceedings of the 51st Annual Meeting of the Association for Computational Linguist (ACL), 2013.

[2] Wang,Meng qiu and Manning,Christopher. Probabilistic tree-edit models with structured latent variables for textual entailment and question answering. The Proceedings of the 23rd International Conference on Computational Linguistics (COLING), 2010.

[3] Wang, Mengqiu, Smith, Noah, and Teruko, Mitamura. What is the jeopardy model? a quasisynchronous grammar for qa. The Proceedings of EMNLP-CoNLL, 2007.

[4] Heilman, Michael and Smith, Noah A. Tree edit models for recognizing textual entailments, paraphrases, and answers to questions. Annual Conference of the North American Chapter of the Association for Computational Linguistics. Association for Computational Linguistics

(NAACL), 2010.
[5] Tomas Mikolov, Ilya Sutskever, Kai Chen, Greg S Corrado, and Jeff Dean, "Distributed representations of words and phrases and their compositionality," in Advances in Neural Information Processing Systems 26, pp. 3111–19. Curran Associates, Inc., 2013.
[6] Hochreiter, Sepp and Schmidhuber, Jurgen. Long short-term memory. Neural Computation, 1997.
[7] Yoon Kim, "Convolutionalneural networksfor sentence classification," in Proceedings of the 2014 Conference on Empirical Methods in Natural Language Processing (EMNLP), Doha, Qatar, October 2014, pp. 1746–51, Association for Computational Linguistics.
[8] Feng, Minwei, Xiang, Bing, Glass, Michael, Wang, Lidan, and Zhou, Bowen. Applying deep learning to answer selection: A study and an open task. IEEE Automatic Speech Recognition and Understanding Workshop (ASRU), 2015.
[9] Ming Tan, Cicero dos Santos, Bing Xiang & Bowen Zhou. Lstm-based deep learning models for non-factoid answer selection. arXiv preprint arXiv:1511.04108, 2015.
[10] Wang, Di and Nyberg, Eric. A long short-term memory model for answer sentence selection in question answering. Proceedings of the 53rd Annual Meeting of the Association for Computational Linguistics and the 7th International Joint Conference on Natural Language Processing, 2015.
[11] Bahdanau, Dzmitry, Cho, KyungHyun, and Bengio, Yoshua. Neural machine translation by jointly learning to align and translate. Proceedings of International conference of learning representations, 2015.
[12] Tomas Mikolov, Ilya Sutskever, Kai Chen, Greg S Corrado, and Jeff Dean. Distributed representations of words and phrases and their compositionality. Advances in neural information processing systems. 2013: 3111-19.
[13] Graves, Alex, Mohamed, Abdel-rahman, and Hinton, Geoffrey. Speech recognition with deep recurrent neural networks. In IEEE International Conference on Acoustics, Speech and Signal Processing (ICASSP), 2013.

IOP Conf. Series: Materials Science and Engineering **323** (2018) 012024 doi:10.1088/1757-899X/323/1/012024

Research on filter's parameter selection based on PROMETHEE method

Hui-min ZHU[1], Hang-yu WANG[2,*] and Shi-yan SUN[1]

[1]College of Weaponry Engineering, Naval University of Engineering, Wuhan, Hubei, 430033, China
[2]Naval University of Engineering, Wuhan, Hubei, 430033, China.

*E-mail: 415745291@163.com

Abstract. The selection of filter's parameters in target recognition was studied in this paper. The PROMETHEE method was applied to the optimization problem of Gabor filter parameters decision, the correspondence model of the elemental relation between two methods was established. The author took the identification of military target as an example, problem about the filter's parameter decision was simulated and calculated by PROMETHEE. The result showed that using PROMETHEE method for the selection of filter's parameters was more scientific. The human disturbance caused by the experts method and empirical method could be avoided by this way. The method can provide reference for the parameter configuration scheme decision of the filter.

1. Introduction

The Gabor function could simulate the working mechanism of biology visual neuron. The target with a certain shape in the image can be accurately recognized only with a series of proper filter's parameters. The preference ranking organization methods for enrichment evaluations (PROMETHEE) is an important multi attributes decision making method, which can be used to solve the problem about image filtering parameter selection.

PROMETHEE is a series of methods based on Outranking Relation (OR) and Outranking Methods (OM). It is a classic multi attributes decision method, and people use this method to solve many problems. Jiang used the PROMETHEE method to research the multiple attribute decision making problems in [1]. SUN researched and compared the robustness of multiple attribute decision making methods, and pointed out in several articles in [2-4]. Dong and other people used PROMETHEE method to simulate some specific problems, proved that this method can provide a good reference for solving incomplete information decision-making and their research are described in the published papers [5-10]. Fang provided a weight calculation method for multiple attribute decision making problems in [11].

Gabor filter has been used in computer vision widely, which meets the demand of image processing during the battlefield reconnaissance. Gabor filter is used in various fields of image processing widely. Chen combines Gabor filter and ant colony algorithm for segmented images, he shows his research in [12]. Guo researched the adaptive image registration based on Gabor filter in [13]. Chen completed the combination of Gabor filter and ICA technology for texture classification in [14]. Ye and Song else used the Log-Gabor filter and its improved method for researching the face recognition in [15] and [16]. Tao used Gabor filter to recognize the road from high resolution remote sensing images in [17].

It's the key to select the correct filter's parameters for sifting out the distracters in the image, which is important for accurately identifying the targets.The principle of Gabor filter is similar to the biological optic nerve system, and the accurate target recognition of the image is beneficial to the decision maker's analysis. The key to solve the target recognition of battlefield image is the configuration of filter's parameters. The selection of the parameters in the Gabor filter function is regarded as the criterion in the PROMETHEE method. The selection of parameters can be solved by this way, and the optimal scheme could be selected according to different requirements.

2. Model establishments

2.1. Objective function

The basic expression of two dimensional Gabor function is like:

$$\Psi(u,v,\sigma_G,\beta_G,\omega_0)=\frac{1}{2\pi\sigma_G^2\beta_G^2}\cdot e^{\left[-\frac{1}{2}\left(\frac{u^2}{\sigma_G^2}+\frac{v^2}{\beta_G^2}\right)\right]}\cdot e^{\left[2\pi i\omega_0(u+v)\right]} \tag{1}$$

In this function, (u,v) means a set of data signal, ω_0 means the spatial frequency domain of complex planar wave. The λ_G means the wavelength of filter, that is the scale of Gabor filter, and $\lambda_G \geq 2$. The σ_G and β_G represent the Gauss standard deviation in the direction of u and v respectively. And the i represents the imaginary part of the filter function. We assume that the scales are the same in u and v directions, $\sigma_G = \beta_G$. In order to remain the shape and scale of the filter unchanged during the filtering, it's necessary to specify a scale factor k on the wavelength of the filter, which is controlled in the x axial direction and the y axis direction, and $\sigma_G = \frac{\lambda_G}{k}$.

Therefore, the filter is expressed as:

$$\Psi(u,v,\lambda_G/k,\omega_0)=\frac{k^2}{2\pi\lambda_G^2}\cdot e^{\left[-\frac{1}{2}\left(\frac{(u^2+v^2)k^2}{\lambda_G^2}\right)\right]}\cdot e^{\left[\frac{2\pi i\omega_0(u+v)}{\lambda_G}\right]} \tag{2}$$

When the direction of the Gabor filter is φ_G , the can be rewritten as:

$$\Psi(u,v,\lambda_G/k,\varphi_G)=\frac{k^2}{2\pi\lambda_G^2}\cdot e^{\left(-\frac{(u^2+v^2)k^2}{2\lambda_G^2}\right)}\cdot e^{\left[\frac{2\pi i(u\cos\varphi_G+v\sin\varphi_G)}{\lambda_G}\right]} \tag{3}$$

Therefore, the Gabor function may be decomposed into two functions like so that the scale and displacement of the target could be judged.

$$\begin{cases}\Psi_R(u,v,\lambda_G/k,,\varphi_G)=\frac{k^2}{2\pi\lambda_G^2}e^{\left[-\left(\frac{(u^2+v^2)k^2}{\lambda_G^2}\right)\right]}\cos\left(\frac{2\pi(u\cos\varphi_G+v\sin\varphi_G)}{\lambda_G}\right) & \textit{Real part}\\ \Psi_I(u,v,\lambda_G/k,\lambda_G/k,\varphi_G)=\frac{k^2}{2\pi\lambda_G^2}e^{\left[-\left(\frac{(u^2+v^2)k^2}{\lambda_G^2}\right)\right]}\sin\left(\frac{2\pi(u\cos\varphi_G+v\sin\varphi_G)}{\lambda_G}\right) & \textit{Imaginary part}\end{cases} \tag{4}$$

2.2. Model of Multi attribute decision making

PROMETHEE provides priority for decision makers in the case of multi attribute and incomplete information. The database about the attributes of different targets is established, and marks this as the criterion set $\mathbf{J}(j_1,j_2,\cdots j_x,\cdots,j_m)$. According to, the parameters need to be set for Gabor filtering include $(\lambda_G, k, \varphi_G)$. The number of effective parameter combination schemes is n , so set the scheme $\mathbf{S}(s_1, s_2, \cdots, s_y, \cdots, s_n)$. According to the preference of decision maker, set the weight coefficient about every target $\boldsymbol{\omega}(\omega_1, \omega_2, \cdots, \omega_x, \cdots, \omega_m)$. Where $\sum_{i=1}^{m} \omega_i = 1$, and ω were greater than 0.

Establishing the priority function as $p_{j_x}(s_i, s_k)$, this function represents the priority of the scheme s_i relative to the scheme s_k on the criterion j_x, and the form of the priority function can be referenced in the Ref.[5]. The priority of each scheme is calculated by inflow and outflow, and the priority index between the two schemes is obtained, such as:

$$d_{j_x}(s_i, s_k) = \frac{\sum_{x=1}^{m} \omega_x \cdot p_{j_x}(s_i, s_k)}{\sum_{x=1}^{n} \omega_x}; d_{j_x}(s_i, s_i) = 0 \tag{5}$$

The is expressing the corresponding preference weight ω_x to the attribute j_x has been given by the decision makers in the case of incomplete information. The priority between the scheme s_i and the scheme s_k is being calculated for comparing the priority intensity of these two schemes under this attribute.

The definition of s_y net flow is:

$$\begin{cases} \phi^{+}(s_y) = \dfrac{1}{n-1} \sum\limits_{k=1, k \neq y}^{n} d_{j_x}(s_y, s_k) \\ \phi^{-}(s_y) = \dfrac{1}{n-1} \sum\limits_{k=1, k \neq y}^{n} d_{j_x}(s_k, s_y) \end{cases} \tag{6}$$

The priority relation of PROMETHEE is determined by:

$$\phi(s_y) = \phi^{+}(s_y) - \phi^{-}(s_y) \tag{7}$$

where, the $\phi_{j_x}(s_y)(j_x \in \mathbf{J}, s_y \in \mathbf{S})$ satisfied the following.

(1) $-1 \leq \mathbf{\Phi}_{j_x}(s_y) \leq 1$

(2) $\sum_{x=1}^{m} \omega_x \mathbf{\Phi}_{j_x}(s_y) = \mathbf{\Phi}(s_y)$

(3) $\sum_{y=1}^{n} \phi_{j_x}(s_y) = 0$

And.

$$\mathbf{\Phi}_{j_x}(s_y) = \begin{bmatrix} \phi_{j_1}(s_1) & \cdots & \phi_{j_2}(s_1) & \cdots & \phi_{j_m}(s_1) \\ \phi_{j_1}(s_2) & & \phi_{j_2}(s_2) & & \phi_{j_m}(s_2) \\ \vdots & & & & \\ \phi_{j_1}(s_y) & \cdots & \phi_{j_2}(s_y) & \cdots & \phi_{j_m}(s_i) \\ \phi_{j_1}(s_n) & \cdots & \phi_{j_2}(s_n) & \cdots & \phi_{j_m}(s_n) \end{bmatrix} \tag{8}$$

It represents the entire priority index net of all schemes based on different attribute. So we can get the collection of all schemes. When the value of $\phi(s_y)$ is bigger, the priority of the scheme s_y is higher.

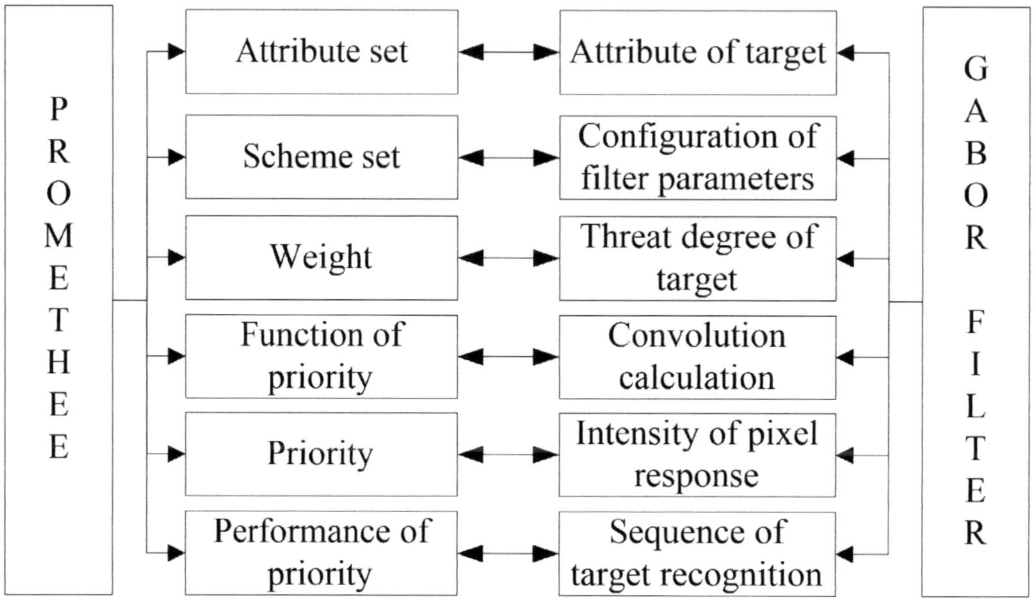

Figure 1. The element correspondence of two methods.

The sets involved in PROMETHEE method include the attribute set and the scheme set, such as figure 1. Assign weight values to each attribute and select an appropriate priority function model. The complete order of scheme priority can be calculated through the comparison between the priority schemes and the calculation of each scheme net.

It is necessary to analyze the target attributes to deal with the filtering of image. The target model database about to different military targets need to be established. The "attribute" set is formed by summing up the appearance, size, color and other characteristics. The different filtering parameters are selected form the scheme $s_y = (\lambda_G, k_x, \varphi_G)$. In order to construct target firing sequence chain, the threat degree of target need to be analyzed, the normalized matrix of index feature need to be established, the weight set of target attribute could be formed. And the firing sequence corresponds to the weight value.

The response value of the pixel is obtained through the convolution calculation in the image. The different parameter schemes could obtain the different pixel response value. Which the response value is higher, the priority of the target recognition is higher. The priority of target recognition is more consistent to the sequence of target threat degree, the scheme is more suitable.

3. Image processing based on Gabor filtering

3.1. Parameters of Gabor filter

The parameters of the Gabor filter include $(\lambda_G, k, \varphi_G)$. When the decision is not made by using the PROMETHEE method, the parameters of the filter can only be set for many times to distinguish the battlefield target in image. The better recognition result and the filter's parameter combination scheme are selected from the multi group of filter results.

The wavelength is selected as an example in this paper, so, the scale factor k is set to 0.5, and the filtering direction φ_G is set to $\frac{\pi}{4}$ and $\frac{3\pi}{4}$. We get the best filtering result when $\lambda_G = 2 \text{ or } 3$ without any decision algorithm intervention. The original image is shown in figure 2(a), set $\varphi_G = \frac{3\pi}{4}$ first, and the filtering result when $\lambda_G = 2$ and $\lambda_G = 3$ is shown in figure 2(b) and (c).

IOP Conf. Series: Materials Science and Engineering **323** (2018) 012024 doi:10.1088/1757-899X/323/1/012024

(a) Original image (b) Filtering result when $\lambda_G = 2$ (c) Filtering result when $\lambda_G = 3$

Figure 2. Image to be filtered.

In the figure 2 (a), the target number 1 and 2 are tank models, the target number 3-6 are aircraft models parked on the airport pavement and the military airports as a building target, numbered 7.

The texture features of tank model and aircraft model are different from the surrounding environment in military image. The airport pavement shows line characteristics obvious. Every pixel point will respond in different direction after Gabor filtering. It can be considered that the pixels are likely to be the target location or the contour of the target area when these pixel points have large texture response value. The pixel response value can be getting by:

$$T(u,z) = P(u,z) * \Psi_R(u,v,\lambda_G / k, \varphi_G) \tag{9}$$

where, $T(u,z)$ represents the pixel response value, $P(u,z)$ represents the gray value of original image, and '*' represents convolution.

3.2. Analysis of image filtering results

In order to recognize the targets, it is necessary to analyze the gray histogram of each filtering result. The peak value in the gray histogram is g_{max} . T_g represents the number of the pixel which the gray value is in the range of $g_{max} \pm r$, r is 15. The image processing can be judged successful when T_g is greater than D_{max} , and the desired target may be identified. Otherwise, the desired target cannot be accurately identified from the image.

The pixels which their response value are in the range of $g_{max} \pm 15$ when $\lambda_G = 2$ and 3 are displayed, the results and the corresponding gray histogram are show in figure 3.

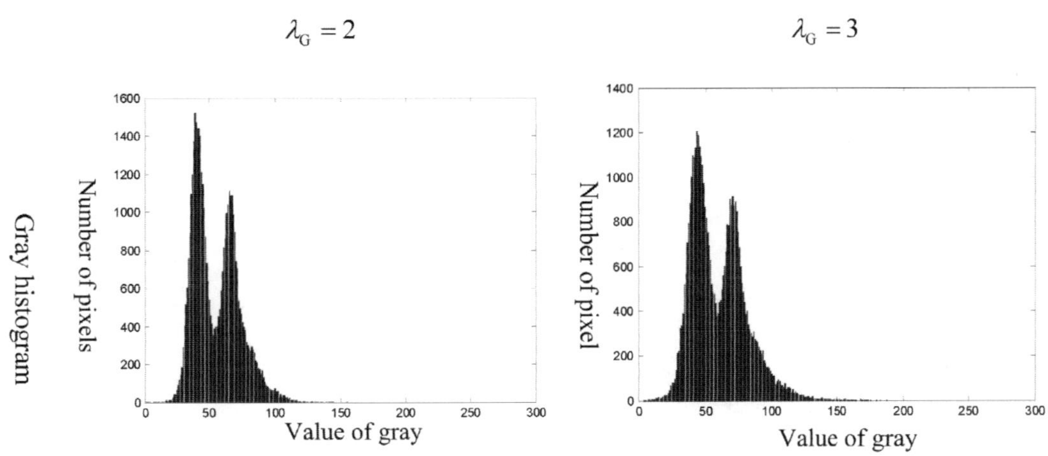

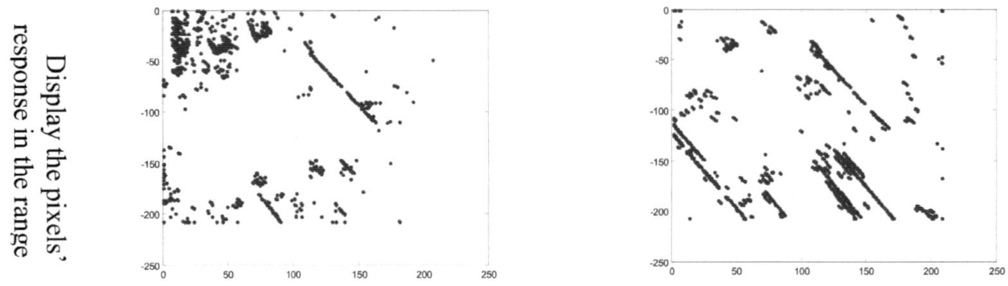

Figure 3. The result of filtering.

The same method is used to calculate when $\varphi_G = \frac{\pi}{4}$, and the results of the two image processing are combined finally in figure 4.

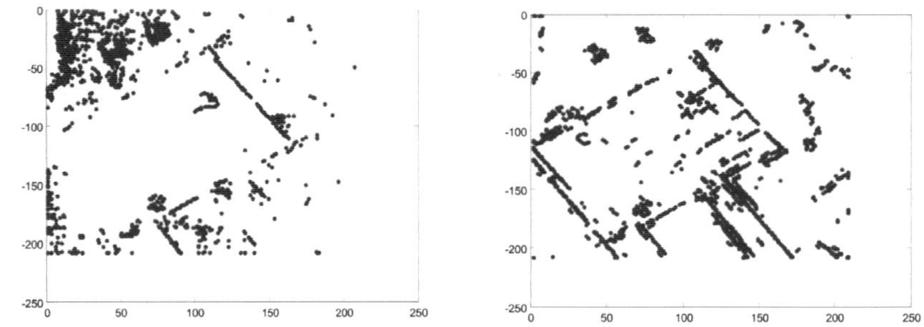

(a) $\lambda_G = 2, \varphi_G = \frac{\pi}{4}, \varphi_G = \frac{3\pi}{4}$ (b) $\lambda_G = 3, \varphi_G = \frac{\pi}{4}, \varphi_G = \frac{3\pi}{4}$

Figure 4. The pixel response of multi angle and different wavelength.

The two results can be evaluated only by visual, and there is no scientific evaluation basis to determine whether the choice of parameters is appropriately. The filtering effect is better from the clarity of the graph when $\lambda_G = 3$.

4. Gabor filter's parameter scheme decision making based on PROMETHEE

4.1. Priority function

There are 7 targets in the attribute set from figure 2(a). The wavelength varies from 1-8, so, there are 8 options in the scheme set, that means $\mathbf{S}(s_x)=\mathbf{S}(s_1(\lambda_1=1),\ s_2(\lambda_2=2),\cdots s_8(\lambda_8=8))$. The priority function can be chosen as in this case.

$$p(T_g) = \begin{cases} 0 & T_g \le D_{\min} \\ \dfrac{(T_g - D_{\min})}{(D_{\max} - D_{\min})} & D_{\min} < T_g \le D_{\max} \\ 1 & T_g > D_{\max} \end{cases} \qquad (10)$$

4.2. Weight

The threat degree of target is taken as the weight of the PROMETHEE method according to figure 1. In order to obtain the weight of every attribute in the decision problem accurately, the target assessment will be carried out in the following section.

According to the target type, size, state and direction (fairway crosscut) of movement, tactical threat, attack ability and so on, using the method of catastrophe decision model, the target threat assessment model is set up as shown in figure 5.

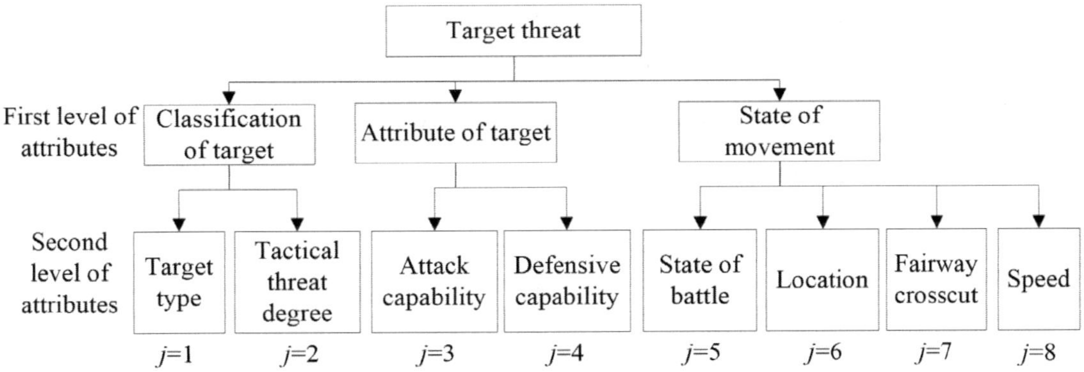

Figure 5. Target threat assessment index classification.

Assigning values to the target attributes as shown in table 1.

Table 1. Threat assessment index of target.

Attributes \ Target number	No.1	No.2	No.3	No.4	No.5	No.6	No.7
Target type	3	3	6	6	6	6	8
Tactical threat degree	3	3	7	4	7	4	8
Attack capability	5	5	6	1	6	1	1
Defensive capability	4	4	5	1	5	1	3
State of battle	5	5	6	3	6	3	3
Location	4	4	5	2	5	2	8
Fairway crosscut	1	1	7	2	7	2	3
Speed	3	3	2	2	2	2	1

The threat attributes of each target are represented by $TV_{\mathrm{P}_i,j}(i=1,2,\cdots,n;j=1,2,\cdots,7)$, where P_i is the target number, and j is the number of second level of attributes. The standardization index value of each target can be expressed as:

$$\begin{aligned} N\_value_{\mathrm{P}_i j} &= \frac{TV_{\mathrm{P}_i j} - TV_j(\min)}{TV_j(\max) - TV_j(\min)}, \quad TV_{\mathrm{P}_i j} \neq TV_j(\min) \\ N\_value_{\mathrm{P}_i j} &= \frac{TV_j(\max) - TV_{\mathrm{P}_i j}}{TV_j(\max) - TV_j(\min)}, \quad TV_{\mathrm{P}_i j} = TV_j(\min) \end{aligned} \tag{11}$$

According to the importance of attributes, the standardization index values are normalized as:

$$\begin{cases} No_{\mathrm{P}_i 1} = \sqrt{N\_value_{\mathrm{P}_i 1}}; No_{\mathrm{P}_i 2} = \sqrt[3]{N\_value_{\mathrm{P}_i 2}}; No_{\mathrm{P}_i 3} = \sqrt{N\_value_{\mathrm{P}_i 3}}; No_{\mathrm{P}_i 4} = \sqrt[3]{N\_value_{\mathrm{P}_i 4}}; \\ No_{\mathrm{P}_i 5} = \sqrt{N\_value_{\mathrm{P}_i 5}}; No_{\mathrm{P}_i 6} = \sqrt[3]{N\_value_{\mathrm{P}_i 6}}; No_{\mathrm{P}_i 7} = \sqrt[4]{N\_value_{\mathrm{P}_i 7}}; No_{\mathrm{P}_i 7} = \sqrt[5]{N\_value_{\mathrm{P}_i 8}} \end{cases} \tag{12}$$

The normalization values of second level attributes are calculated to complement their first level attributes. The first level attributes value of each target are summarized as:

$$\begin{cases} AV_{\mathrm{P}_i,1} = \min(No_{\mathrm{P}_i,1}, No_{\mathrm{P}_i,2}) \\ AV_{\mathrm{P}_i,1} = \max(No_{\mathrm{P}_i,3}, No_{\mathrm{P}_i,4}) \\ AV_{\mathrm{P}_i,1} = \dfrac{No_{\mathrm{P}_i,5} + No_{\mathrm{P}_i,6} + No_{\mathrm{P}_i,7} + No_{\mathrm{P}_i,8}}{4} \end{cases} \tag{13}$$

Thus, the threat degree of the target is:

$$TV_{P_1}=0.800628;TV_{P_2}=0.800628;TV_{P_3}=0.82444;TV_{P_4}=0.498709;$$
$$TV_{P_5}=0.82444;TV_{P_6}=0.498709;TV_{P_7}=0.741681; \quad (14)$$

The weight of PROMETHEE can be expressed as by normalization,

$$\omega_{P_1}=0.1448;\omega_{P_2}=0.1448;\omega_{P_3}=0.2172;\omega_{P_4}=0.0788;\omega_{P_5}=0.2172;\omega_{P_6}=0.078\,8;\omega_{P_7}=0.1182 \quad (15)$$

Therefore, the most threatening targets are the number 3 and number 5 target.

4.3. Scheme decision

According to the Section B in ChapterII,we can calculate the net flow of each scheme $\phi(s_y)$.

$$\phi(s_1)=0.4836;\phi(s_2)=0.6626;\phi(s_3)=0.5440;\phi(s_4)=0.2726;$$
$$\phi(s_5)=-0.0718;\phi(s_6)=-0.4145;\phi(s_7)=-0.6518;\phi(s_8)=-0.8247 \quad (16)$$

The priority of the scheme is:

$$a_2 \succ a_3 \succ a_1 \succ a_4 \succ a_5 \succ a_6 \succ a_7 \succ a_8 \quad (17)$$

Therefore, the optimal scheme is $s_2(\lambda_2=2)$.

4.4. Verification

The verification method is to analyze the results of two image processing results when $\lambda_2=2$ and $\lambda_3=3$. The higher the proportion of pixels satisfying the response value range, the topper of the target recognition order is. The order of target recognition should be corresponding to their threat degree sequence. The calculation results of the pixel response values of different targets in the two filtering cases are such as:

$$\lambda=2(Ratio\_1=96.86\%;Ratio\_2=94.53\%;Ratio\_3=97.35\%;Ratio\_4=92.75\%;$$
$$Ratio\_5=99.67\%;Ratio\_6=92.50\%;Ratio\_7=54.49\%)$$
$$\lambda=3(Ratio\_1=95.17\%;Ratio\_2=86.72\%;Ratio\_3=92.86\%;Ratio\_4=71.37\%;$$
$$Ratio\_5=93.75\%;Ratio\_6=87.73\%;Ratio\_7=51.69\%) \quad (18)$$

The proportion of pixels satisfying the response range is sorted as follows:

$$\lambda=2:5>3>1>2>6>4>7;\ \lambda=3:1>5>3>6>2>4>7 \quad (19)$$

Compared with the two schemes, the target number 5 and 3 can be identified first when $\lambda=2$ .It is more consistent with the target threat sequence.

5. Conclusions

Target recognition determines the success or failure of military operations. Multi-targets appearing in the battlefield image is very common. The Gabor filter is used for solving the problem of target recognition, and the selection of filter's parameters is the multi-attribute decision-making problem faced by the decision makers. And then, the PROMETHEE method is used for parameters selection in this paper, which provides a scientific basis for the selection of filter's parameters. The following conclusions are obtained through the research of this paper.

The combination scheme of filter's parameter is numerous; the proper scheme could help recognize the important or the main target. The filter's parameters can also be selected by expert method. However, the PROMETHEE can provide a reasonable and scientific decision on the parameter scheme choice. The human disturbance caused by subjective factors is avoided.

The PROMETHEE method is applied to the image filtering problem, and the element correspondence model is established, which provides a better filter's parameter selection method.

The feasibility and effectiveness of the PROMETHEE method in solving the decision problem of Gabor filter's parameters are verified by simulations.

References

[1] Jiang Guangtian. Research on Theoretics and Methods for Solving Stochastic Multiple Attribute Decision Making Problems [D]. Shenyang, Northeastern University, 2009 pp 50-70

[2] Sun Shiyan, Qiu Zhiming, Wang Hangyu. PROMETHEE Parameters Inferring Method Based on Robustness Analysis [J]. *Journal of WUT(Information & management engineering)*, 2007,29(6) pp 32-34

[3] Sun Shiyan, Wang Hangyu, Qiu Zhiming. Robustness analysis method of PROMETHEE weights with incomplete information [J]. *Command and Control & Simulation*, 2007,29(1) pp 47-49

[4] Sun Shiyan, Zhu huimin. PROMETHEE's parameters setting method based on robustness analysis [J]. *Systems Engineering and Electronics*, 2017, 39(1) pp 120-24

[5] Dong Fugui , Yuan Dandan, Shang Meimei. Distributed Renewable Generations Planning Scheme Evaluation Based on AHP and PROMETHEE [C]. *2015 Int. Conf. on Simulation, Modeling and Mathematical Statistics (SMMS 2015)*, Chiang Mai, Thailand, 2015 pp 363-68

[6] Wang Jianqiang. PROMETHEE method with incomplete certain information and its application [J]. *Systems Engineering and Electronics*, 2005, 27(11) pp 1909-13

[7] YU Chao, FAN Zhiping. Multiple attribute decision making method with attribute aspirations in the form of random variable [J]. *Control and Decision*, 2015, 30(8) pp 1434-40

[8] LIANG Xia,JIANG Yanping,GAO Meng. Product Selection Methods Based on Online Reviews [J]. *Journal of Northeastern University (Natural Science)*, 2017, 38(1) pp 143-47

[9] Zhao Yehan, Li Xingguo, Yang Ying. R&D project evaluation based on semantic uncertainty PROMETHEE [J]. *Journal of Hefei University of Technology*, 2015, 38(12) pp 1699-705

[10] GUO Yong, QIU Zhiming. The PROMETHEE method in optimizing a number of selection schemes of naval gun weapon system [J]. *Acta Armamentarii*, 1999, 20 (2) pp 97-101

[11] Fang Xiangrong, Zhou Hongan. Approach of uncertain multi-attribute decision-making with preference information of alternatives under partial weight information [J]. *Journal of Xi'an Shiyou University (Natural Science Edition)*, 2011, 26(1) pp 99-105

[12] Chen Jie, Deng Min, Xiao Pengfeng, et al. Texture Segmentation Based on Gabor Filters and Ant Colony Optimization Algorithm [J]. *Geomatics and Information Science of Wuhan University*, 2010, 35(11) pp 1271-74

[13] Guo Xiaojie, Kang Junfang, Jiang Murong. Self-adaptive parameter computing for deformable registration with gabor transformation [J]. *Journal of Yunnan University*, 2009, 31(S1) pp 9-12

[14] Chen Yang, Wang Runsheng. A Method for Texture Classification by Integrating Gabor Filters and ICA [J]. *ACTA ELECTRONICA SINICA*, 2007, 35(2) pp 299-303

[15] Ye Jihua, Lan Qingping, Liu Changhong, et al. Embedded Face Recognition Combined Mutual Information with Log-Gabor Feature [J], *Journal of System Simulation*, 2016,28(9) pp 2214-19

[16] Song Wei, Zhao Qingjie, Song Hong, et al. Keyblock distribution and Gabor filter based facial expression recognition algorithm [J]. *Journal of Central South University (Science and Technology)*, 2013, 44(2) pp 239-43

[17] Tao Chao, Tan Yihua, Peng Bifa, et al. A Probabilistic Latent Semantic Analysis Based Classification for High Resolution Remotely Sensed Imagery [J]. *Acta Geodaetica et Cartographica Sinica*, 2011,40(2) pp 156-62

IOP Conf. Series: Materials Science and Engineering **323** (2018) 012025 doi:10.1088/1757-899X/323/1/012025

Design and development of an IoT-based web application for an intelligent remote SCADA system

Kuang-Chi Kao[1*], Wei-Hua Chieng[1] and Shyr-Long Jeng[2]

[1] National Chiao Tung University, Department of Mechanical Engineering, Hsinchu, Taiwan.
[2] Ta Hwa University of Science and Technology, Department of Electrical and Electronic Engineering, Hsinchu, Taiwan.

*E-mail: chi1010860@gmail.com

Abstract. This paper presents a design of an intelligent remote electrical power supervisory control and data acquisition (SCADA) system based on the Internet of Things (IoT), with Internet Information Services (IIS) for setting up web servers, an ASP.NET model–view–controller (MVC) for establishing a remote electrical power monitoring and control system by using responsive web design (RWD), and a Microsoft SQL Server as the database. With the web browser connected to the Internet, the sensing data is sent to the client by using the TCP/IP protocol, which supports mobile devices with different screen sizes. The users can provide instructions immediately without being present to check the conditions, which considerably reduces labor and time costs. The developed system incorporates a remote measuring function by using a wireless sensor network and utilizes a visual interface to make the human–machine interface (HMI) more instinctive. Moreover, it contains an analog input/output and a basic digital input/output that can be applied to a motor driver and an inverter for integration with a remote SCADA system based on IoT, and thus achieve efficient power management.

1. Introduction

With the rapid development of the Internet in recent years, Internet-based applications such as remote monitoring systems are becoming increasingly popular in industry. Through the boundless Internet, a remote monitoring system allows the user to have remote real-time control of the situation in a factory by using a smartphone or computer.

In the future, factories will have different types of devices, which will need to be integrated in an intelligent manner. Wireless sensor networks (WSNs) are the base technology of the Internet of Things (IoT) [1]. A WSN is a network that uses intelligent sensors to transmit and receive data. The applications of WSN include health, environment, industrial, and traffic monitoring [2][3]. WSNs have not only contributed to the development of IoT [4] but also led to the development of devices and technologies that support the growth of the Internet, such as QR codes, intelligent phones, social networks, and cloud computing [5][6].

The applications of IoT can be divided into four categories, namely transportation and logistics, healthcare, smart environment (home, office, and plant), and personal and social [7]. Power supervision is important for not only factory power management but also home power monitoring [8]. Power line communication requires considerable power monitoring for factories, houses, or vehicles. However, power line communication has drawbacks such as large energy loss [9].

The development of Internet of Vehicles, IoT, and cloud computing technologies is expected to result in increasing demand for intelligent power monitoring systems and remote supervisory control and data acquisition (SCADA) systems [10][11].

Using an IoT-based SCADA system is a convenient and efficient way of remotely monitoring industrial equipment. Application of WSNs or wireless sensor and actuator networks to a SCADA system is beneficial because these networks are based on the well-developed TCP/IP and HTTP protocols for transferring data to a client from a server through a web application [12].

The open database connectivity (OBDC) method can be used for connecting and transferring the data, for example through Microsoft SQL Server. The web server must be established in order to provide a response to the client. Microsoft Internet Information Services (IIS) is a service for setting up the web server and delivering data to the client in Extensible Markup Language (XML) format [13]. In recent years, developers have replaced XML with JavaScript Object Notation (JSON) format to transfer data because JSON is a lightweight data interchange format that is easy for humans to read and write as well as easy for machines to parse and generate. JSON increases the decoding speed of a browser and improves the efficiency of a website. In addition, developers use a model–view–controller (MVC) architectural pattern to design a website because it provides a way to divide a given application into three interconnected parts. The MVC design pattern decouples these major components, enabling efficient code reuse and parallel development. The cloud-based vehicle toll payment system mentioned in [14] adopts the aforementioned strategy for the development of a web application.

A monitoring system can be divided into monitor tier, server tier, cloud tier, and client tier. Different tiers have different functions for using devices and software to achieve remote monitoring and data management [15].

A simple application for a remote monitoring system is a web oscilloscope. The web oscilloscope delivers acquisition data, which is stored in the database via TCP/IP and simulates a real oscilloscope to enable the user to easily observe the signal plot of an electrical circuit [16]. Other applications of remote monitoring include video surveillance, appliance testing, and ocean monitoring [17]. Because several studies have proposed that the graphical interfaces are more acceptable for users than using numerical tables as the interface, all these applications are designed with a friendly human–machine interface to enable users to supervise a situation [18][19].

According to market research of browsers used in intelligent devices [20], people use browsers for browsing websites on devices such as smartphones, tablets, and laptops. Because devices can have different sizes, websites must be designed using responsive web design (RWD) to fit different devices. Therefore, a remote SCADA system is discussed in this paper [21].

2. Concept of the System and Related Technologies

This paper presents the design of a web application based on the ASP.NET framework using Visual Studio 2015. The coding languages used to establish the bridge between the server and client in the MVC structure were C#, SQL, JavaScript, HTML, and CSS. Some popular libraries, such as Razor and jQuery, and a common format for transferring data—JSON—were used in this project to implement the function of Asynchronous JavaScript and XML (AJAX).

2.1. ASP.NET MVC architecture

In this study, an MVC architecture was adopted to design the web application for a remote SCADA system. As shown in figure 1, the developer must divide an application into three types of components:

(a) A model stores data that is then retrieved according to commands from the controllers. It includes business logic and all data used in a project.

(b) A view presents the data from the controller based on changes in the model.

(c) A controller decides the data flow and sends commands to the model and view.

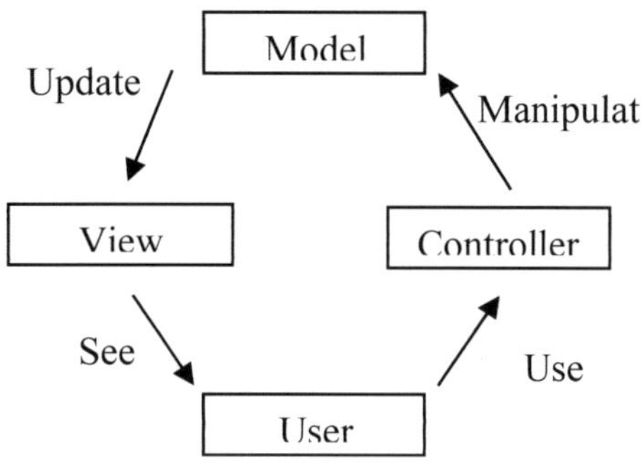

Figure 1. MVC architecture.

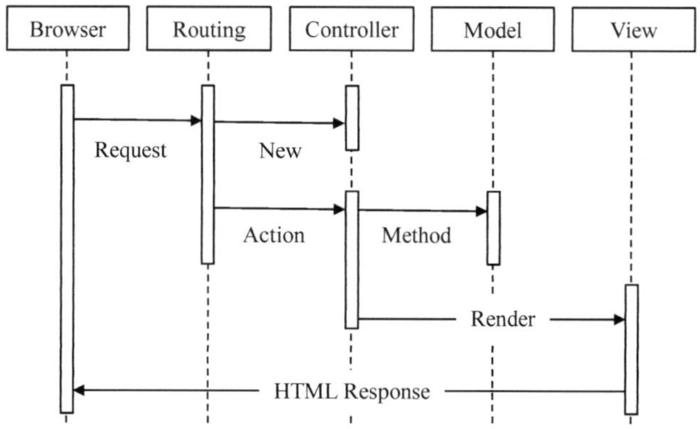

Figure 2. Life cycle of an MVC.

Programming languages such as Java, C#, and PHP have popular MVC frameworks that are used in web application development. In this study, the ASP.NET MVC framework was applied to a remote SCADA system.

The life cycle of an MVC architecture is presented in figure 2. The entry point for every MVC application is URL routing. After the ASP.NET platform receives a request from a browser, a controller determines how it should be handled. The controller then determines the view and presents the view through URL routing.

2.2. Relational database

Microsoft SQL Server 2016 was adopted as a relational database management system to execute four basic functions of a database, namely create, read, update, and delete. Using a database to store data acquired with an analog-to-digital converter and establishing relations among each type of data is a common approach in industrial systems. To connect the web application to an SQL Server, Entity Framework should be used. Entity Framework can help a developer to relate the object in their code with a table in the database. This database technique is called "Code First" because developers initially design the objects and generate the code, and then implement the database. In this study, a view was created that presents the previous data acquired by the system to the users if required.

2.3. RWD

RWD was used in the IoT-based remote SCADA system to develop a friendly user interface. RWD is an approach of web designing aimed at allowing desktop webpages to be viewed in response to the size of the screen. For example, a website with a wide original template will be narrow when the pixels of the device screen are less than 400 px. An RWD website in different devices shows different UI templates. A website designed with RWD adapts the layout to the viewing environment by using fluid, proportion-based grids, and CSS3 media queries. The primary advantage of an RWD website is the reduction in user scrolling and zooming. Reading the data in an RWD website on smart devices makes supervision easier and more accurate than other websites without RWD. Moreover, it removes the problems of incorrectly touching buttons and improves the user scenarios considerably.

2.4. AJAX

Traditional websites must refresh continually to achieve dynamic updating. The users cannot efficiently browse a website because of the time required for waiting for the blank page to refresh before the content of the page appears.

AJAX is a new technique that enables a web application to send and receive data from a server asynchronously. With AJAX, a web page can change content dynamically without needing to reload the entire page. In practice, modern implementations commonly substitute JSON for XML because of the advantages of being native to JavaScript.

AJAX is not a single technology but rather a group of technologies. HTML, CSS, and JavaScript may be used synchronously in a project.

In this study, the AJAX technique was used with the jQuery library in JavaScript to develop the numerical and graphical interfaces for displaying the measurement data. figure 3. (a) shows the practical code for implementing the AJAX function. The red box in figure 3. (b) indicates the area that is dynamically updated with the AJAX technique.

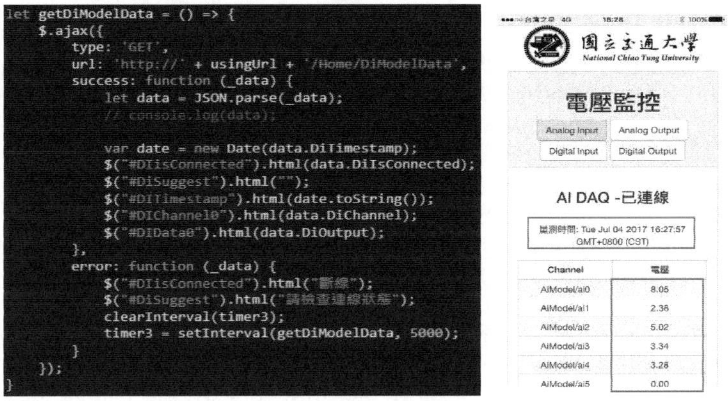

(a) (b)

Figure 3. AJAX monitoring interface of mobile browser: (a) AJAX function with jQuery code. (b) Interface of a smartphone.

2.5. Communication

A system comprising a server PC, an inverter, a WSN module, and a client device was established. Communication between these pieces of hardware is a critical problem. The server PC and inverter are connected through the Modbus protocol and with RS-485 as the electrical interface. Therefore, a USB-RS-485 converter (National Instruments, USA) was used in this study. To transmit data from a Wi-Fi data acquisition (DAQ) device to the server PC, the IEEE 802.11 standard is required. The DAQ device and the server PC can be connected using a Wi-Fi router. Finally, the server PC uses the TCP/IP protocol to connect to the client devices because this system is based on the Internet.

3. Implementation of the IoT-based SCADA System

3.1. Architecture of a remote SCADA system

The system was divided into monitor, server, cloud, and client tiers. The architecture shown in figure 4 shows that the system comprises sensors, a Wi-Fi DAQ device, an inverter, and a motor as the monitor tier; a wireless router and a PC server as the server tier; a database as the cloud tier; and a laptop, a tablet, and a smartphone as the client tier. With this architecture, the system can be practically and structurally implemented and the data flow of the entire system can be confirmed.

3.2. Wireless sensor network with a Wi-Fi DAQ device

This paper presents a structure with a Wi-Fi DAQ device for supervising the voltage and current of the inverter to control the motor. The Wi-Fi DAQ device (cDAQ series; National Instruments) was selected because this device is based on the .NET Framework. Developers can apply this device to their project using a C#-based API and easily connect the application to Microsoft IIS and SQL Server database. Therefore, functions including analog input/output and digital input/output using the browser are presented.

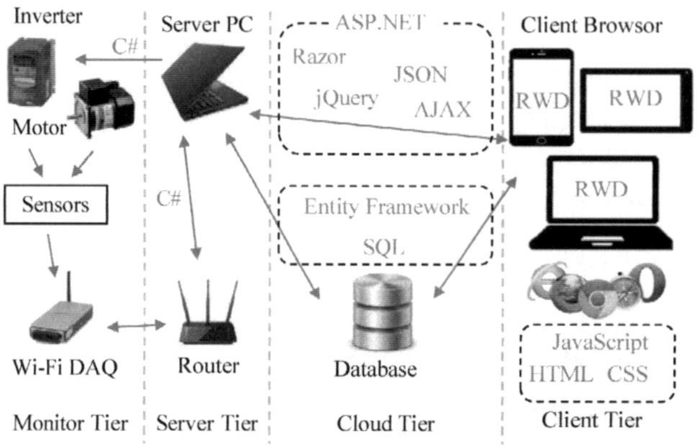

Figure 4. Architecture of a remote monitoring system.

3.3. *Hardware structure of the SCADA system*

The hardware structure of the system is as shown in figure 5. The structure can be simply divided into two parts, namely the wireless monitor and inverter control. The components of the wireless monitor include the Wi-Fi DAQ device, potential meters as the substitute of the sensors, and a simple experimental circuit for testing the analog input/output and digital input/output. The components of the inverter control include an inverter (Rich tech), a USB-RS485 converter, a 1.5-kW motor, a no-fuse breaker, and a 220-V autotransformer. To present the performance of this system, a laptop was used to connect the Wi-Fi DAQ device and inverter. In addition, it was established as a server and presented the scenario in which users attempt to browse the web application of the remote SCADA system.

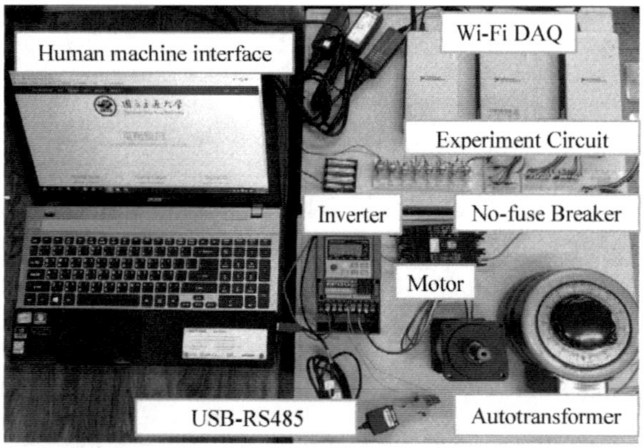

Figure 5. Hardware architecture of the SCADA system.

Figure 6. Data visualization for real-time monitoring.

3.4. *Data visualization*

The data transmission flows in two directions: from database to the client tier and from the monitor tier to the client tier without the database. Both situations are presented using a graphical interface with the AJAX technique to present the data to users.

Without a graphical interface for monitoring, the users cannot read and understand the data easily. Designing a graphical interface for a SCADA system is necessary because it makes users know what happens exactly in the supervision. Compared with the conventional method of using the graphical monitoring interface, AJAX is a much faster approach that does not require refreshing the web page. The advantage is that the data can be visualized considerably more clearly than with a numerical data demonstration. Users can observe the data immediately because data is transmitted to the client directly without waiting for writing and reading of the database. The real-time dynamic monitoring interface is shown in figure 6. The JavaScript library jQuery is used to enable the user read the data more easily and intuitively.

3.5. Remote control of inverter

The system developed in this study not only enables monitoring but also provides remote control. Several basic functions were developed to control the voltage and frequency of the inverter via RS-485 as the physical layer of the Modbus protocol. They can be divided into four main functions which are Analog input (AI), Analog output (AO), Digital input (DI), Digital output (DO) and shown in figure 7.

This paper presents several simple functions of remote control to demonstrate how users can control a system using their smartphones. Using the web application presented in this paper, users can send commands in case of any emergency from wherever they are without being actually present at the location. They can also use any intelligent device that can connect to a wireless network.

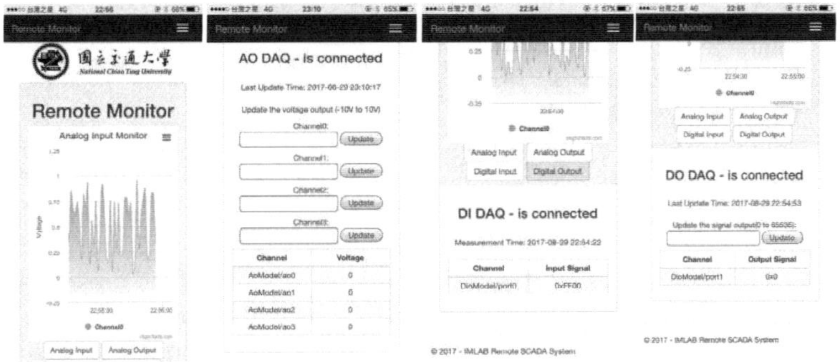

(a) (b) (c) (d)

Figure 7. Implementation analog input/output and digital input/output functions with a smart phone.

4. Conclusions

With the development of IoT, the need for a bridge between things and the Internet is growing. This paper presents a practical IoT-based web application for an intelligent remote electrical power SCADA system. It includes a wireless sensor network for acquiring signal data and receiving information from things that people are interested in to the world through the Internet. This system enables communication with intelligent devices. Users can remotely control and monitor the system by using a browser on their PC or intelligent device.

This system was applied for remote inverter control and monitoring with basic functions. Although the system is incomplete, it has substantial potential for incorporating more functions. This system can be applied in transportation vehicles such as intelligent yachts, autonomous cars, and personal rapid transit and supervise power usage in the future. The measurement data is stored in the database to achieve efficient power management.

References

[1] L. D. Xu, W. He and S. Li, "Internet of Things in Industries: A Survey," in IEEE Transactions on Industrial Informatics, vol. **10**, no. 4, pp. 2233-43, Nov. 2014.

[2] S. Li, L. Xu, and X. Wang, "Compressed sensing signal and data acquisition in wireless sensor networks and internet of things," IEEE Trans. Ind. Informat. vol. **9**, no. 4, pp. 2177–86, Nov. 2013.3

[3] W. He and L. Xu, "Integration of distributed enterprise applications: A survey," IEEE Trans. Ind. Informat. vol. **10**, no. 1, pp. 35–42, Feb. 2014.

[4] L. Wang, L. Xu, Z. Bi, and Y. Xu, "Data filtering for RFID and WSN integration," IEEE Trans. Ind. Informat. vol. **10**, no. 1, pp. 408–18, Feb. 2014.
[5] F. Tao, Y. Laili, L. Xu, and L. Zhang, "FC-PACO-RM: A parallel method for service composition optimal-selection in cloud manufacturing system," IEEE Trans. Ind. Informat., vol. **9**, no. 4, pp. 2023–33, Nov. 2013.
[6] D. Bandyopadhyay and J. Sen, "Internet of things: Applications and challenges in technology and standardization," Wireless Pers. Commun., vol. **58**, no. 1, pp. 49–69, 2011.
[7] L. Atzori, A. Iera, and G. Morabito, "The internet of things: A survey," Comput. Netw. vol. **54**, no. 15, pp. 2787–805, 2010.
[8] L. M. L. Oliveira, J. Reis, J. J. P. C. Rodrigues, A. F. de Sousa, "IoT based solution for home power energy monitoring and actuating," IEEE, 13th Int. Conf. Industrial Informatics (INDIN), Jul. 2015.
[9] Y. Huo, W. Tu, Z. Sheng and C.M. Leung, "A survey of in-vehicle communications: Requirements, solutions and opportunities in IoT," IEEE Conf. 2nd World Forum on Internet of Things (WF-IoT), pp.132–37, 2015.
[10] A. Fachechi, L. Mainetti, L. Palano, L. Patrono, M. L. Stefanizzi, R. Vergallo, P. Chu, R. Gadh, "A new vehicle-to-grid system for battery charging exploiting IoT protocols," IEEE Int. Conf. Industrial Technology (ICIT), 2015.
[11] Z. Ding, B. Yang, R. H. Guting and Y. Li, "Network-Matched Trajectory-Based Moving-Object Database: Models and Applications," IEEE Trans. Intell. Transp. Syst., vol. **16**, no. 4, pp.1918–28, Aug. 2015.
[12] A. M. Grilo, J. Chen, M. Diaz, D. Garrido, A. Casaca, "An Integrated WSAN and SCADA System for Monitoring a Critical Infrastructure," IEEE Transactions on Industrial Informatics, vol. **10,** no. 3, Aug. 2014.
[13] A. Lipnickas, R. Rutkauskas, R. Cerkauskas, "A resource oriented architecture for Web-integrated SCADA applications," IEEE World Conference on Factory Communication Systems (WFCS), May. 2015.
[14] B. Cvijic, D. Bundalo, D. Pasalic, Z. Bundalo, "Cloud based web application supporting vehicle toll payment system," IEEE, 5th Mediterranean Conference on Embedded Computing (MECO), Jun. 2016
[15] E. Kanagaraj, L. M. Kamarudin, A. Zakaria, R. Gunasagaran, A. Y. M. Shakaff, "Cloud-based Remote Environmental Monitoring System with Distributed WSN Weather Stations," IEEE Sensors, Nov. 2015.
[16] R. Marques, J. Rocha, S. Rafael, J. F. Martins, "Design and Implementation of a Reconfigurable Remote Laboratory, Using Oscilloscope or PLC Network for WWW Access", IEEE Transactions on Industrial Electronics, vol. **55**, no. 6, Jun. 2008.
[17] C. Liu, Z. Guo, Y. Feng, F. Hong, W. Jing, "CPCA: The Cloud Platform of Complex Virtual Instruments System Architecture," IEEE Access, vol. **5**, pp. 4350-4360, Mar. 2017.
[18] P. Chynal, J. Sobecki, "Eyetracking Evaluation of Different Chart Types Used for Web-Based System Data Visualization," IEEE Conf. Network Intelligence Conference (ENIC), Jul. 2016.
[19] P. Radunovic, T. Vujicic, A. Balota, "Web application for lightning activity monitoring system (LAMS)," IEEE Conf. Information System and Techologies (CISTI), Jun. 2017.
[20] W. Peng, Y. Zhou, "The Design and Research of Responsive Web Supporting Mobile Learning Devices," IEEE Conf. Educational Technology (ISET), Jul. 2015.
[21] M. A. Moyeen, G. G. Md. Nawaz Ali, Peter Han Joo Chong, "An automatic layout faults detection technique in responsive web pages considering JavaScript defined dynamic layouts," IEEE Conf. Electrical Engineering and Information Communication Technology (ICEEICT), Sep. 2016.

IOP Conf. Series: Materials Science and Engineering **323** (2018) 012026 doi:10.1088/1757-899X/323/1/012026

A study on obstacle detection method of the frontal view using a camera on highway

Van-Quang Nguyen[1], Jeonghyeon Park[2], Changjun Seo[2], Heungseob Kim[1] and Kwangsuck Boo[1,*]

[1]Department of Mechanical and Automotive Engineering, Inje University, Gimhae, Gyeongnam, Korea.
[2]Department of Electronic Engineering, Inje University, Gimhae, Gyeongnam, Korea.

*Email: mechboo@inje.ac.kr

Abstract. In this work, we introduce an approach to detect vehicles for driver assistance, or warning system. For driver assistance system, it must detect both lanes (left and right side lane), and discover vehicles ahead of the test vehicle. Therefore, in this study, we use a camera, it is installed on the windscreen of the test vehicle. Images from the camera are used to detect three lanes, and detect multiple vehicles. In lane detection, line detection and vanishing point estimation are used. For the vehicle detection, we combine the horizontal and vertical edge detection, the horizontal edge is used to detect the vehicle candidates, and then the vertical edge detection is used to verify the vehicle candidates. The proposed algorithm works with of 480×640 image frame resolution. The system was tested on the highway in Korea.

1. Introduction

According to the WHO, each year lives of approximately 1.25 million people are cut short as a result of road traffic crash. Between 20 and 50 million people suffer from non-fatal injuries, which sometimes incur disabilities [1]. Road traffic injuries bring considerable economic losses to victims, their families, and nations as a whole. Therefore, in 2016 many firms or corporation has declared that they were and will participate in the development of the automatic vehicle. Volvo Corporation has promised that by 2020, no one will be killed or seriously injured by one of its new cars by using driving assistance system and warning [2].

Many researchers have studied and proposed the different strategies and track vehicle and lane, but most of the works have been done separately. For the lane detection, there are many main features that can extract the lane mark pixel from an image, consisting of edge feature [3], color feature [4], Haar-like feature [5], and hough transform [6]. Among the vehicle detection approaches, there are also many approach methods based on features, such as Haar-like Feature [7], Haar-like feature and artificial neuron networks [8], Histogram of oriented gradients (HOG) and support vector machine (SVM) classifier [9, 10] proposed. These proposed algorithms, which are powerful and robust, operate in real-time with high accuracy. However, only road detection or vehicle detection is not sufficient to support driving assistance system, warning system, or lane change system.

In this paper, the key idea is to integrate the relationship between lanes and vehicles to support the warning system, and a driver assistance system. In figure 1a, when drivers want to change lane on the right-side, they have to monitor the frontal lane, and right-side lane. For instance, there are vehicles on these lanes and near the driver's vehicle, so drivers could not change lane, inversely there are not vehicles on these lane, drivers could change lane. This is one example for the purpose of my paper. This idea consists of two main steps, the information lane will be detected in the first step. And then, vehicle will be detected inside area among the detected lanes by vehicle features. Figure 1b shows the block diagram of proposed system. In the block diagram, we have a notions as a region of interest (ROI). ROI is defined as an area close to the test vehicle. The full image is used to detect vehicles. There are two phases: 1) lane detection on ROI, and 2) vehicles detection based on the detected lanes. This paper is divided as follows. Section 2 describes steps for the lane detection. The method of vehicle detection and tracking is shown in section 3. Section 4 presents the experimental results.

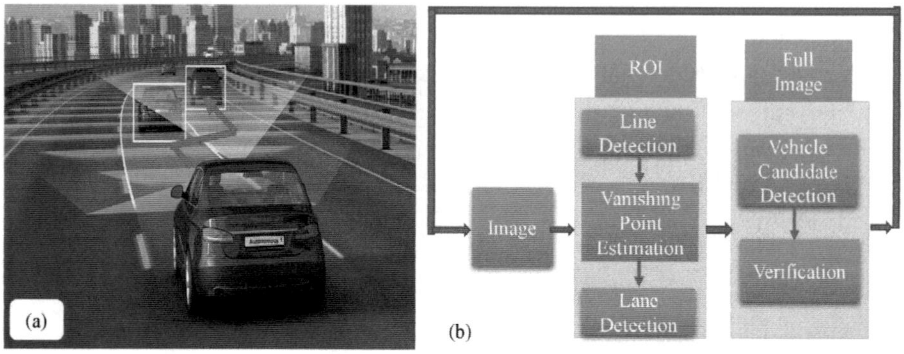

Figure 1. (a) Vehicles monitoring for lane change, (b) The proposed method.

2. Lane detection

2.1. Line analysis

The first step of the method is to discover the line segments from the image inputs. For line detection, EDLines are used [11], EDlines algorithm only takes from 10ms to 20ms with an Intel 2.2 GHz CPU Moreover, EDLines algorithm runs even faster when it is appled to ROI. For any image in this work, we determine a region interest (ROI) for lane detection, ROI are defined as the rectangle inside the red lines, as shown in figure 2a. Figure 2b shows the line segment detected by EDlines.

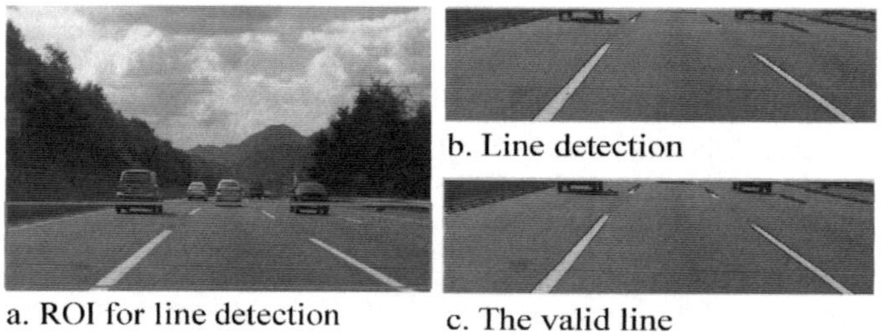

Figure 2. Line detection and analysis.

After line segments are extracted from the image, two eliminative steps are performed to remove the irrelevant line segments. Firstly, all of the vertical line segments, horizontal line segments, and very short line segments are removed and these are the red line segments in figure 2b. The remaining line segments are shown in figure 2c. Next, we have to reject the irrelevant remaining line segments and keep the relevant ones. The relevant line segments are the lane markings, and the lines intersecting the lane markings at the vanishing point. The line segments are divided into two subsets, left and right candidate sets. The left segments are indicated by blue and the right segments are indicated by green. The angles between all of these segments and the horizontal axis are calculated as shown in figure 3a, in which θ_{left} is the angle between left segments and horizontal axis and θ_{right} is the angle between right segments and horizontal axis. We have set range limits for each angle, 30° to 85° for θ_{left} and

IOP Conf. Series: Materials Science and Engineering **323** (2018) 012026 doi:10.1088/1757-899X/323/1/012026

120° to 175° for θ_{right} . It means that any segment that does not belong to the ranges will be removed. In figure 3b, the red segment will be removed and the relevant segments of the two sides will be kept for the next step.

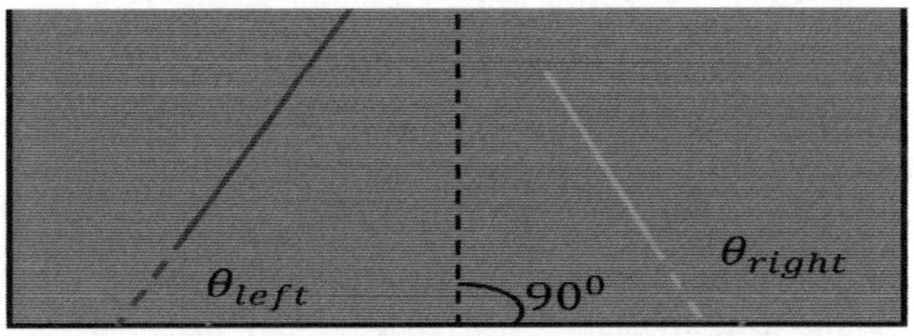

a. Angle of segment with horizontal

b. The selected and eliminated line segments based on their angle

Figure 3. Line segment selection based on their angle with horizontal axis.

2.2. *Vanishing point estimation*

As previously mentioned, we have two clusters of the relevant segments, the left and the right. The vanishing point is the intersection where the lane markings or the road boundaries cross. Therefore, all lines on each cluster are extended as in figure 4a and we have found the intersection points where each line of the left cluster meets the right cluster ones as shown in figure 4b (red points). In order to define the vanishing point, we create a grid formed by square cells of size equal to the permitted error in estimating the vanishing point position. Finally the cell that has the maximum number of the intersection points, is selected and its centre weight is returned as an approximation of the vanishing point position. In figure 4c, we present the test image in which the vanishing point is clearly visible with a red.

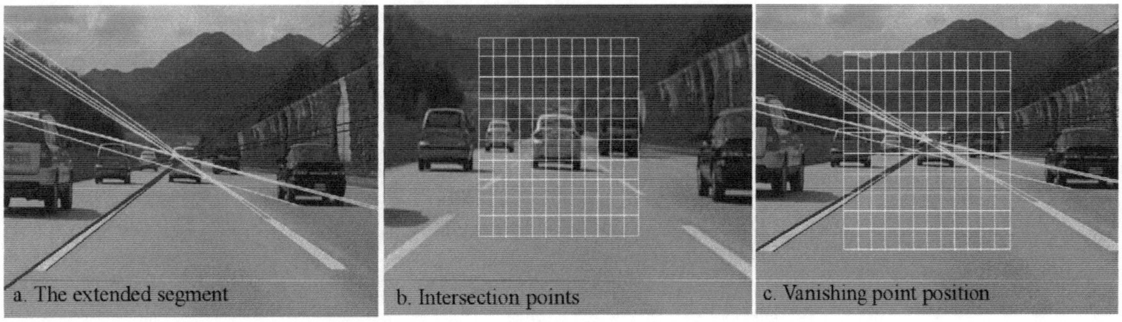

Figure 4. Vanishing point estimation.

2.3. *Lane detection*

To detect lane markings, the horizontal straight line is created, they intersect with the extended segment at red points, these points are in the red circles. Based on the distance among these points, the points are close which are clustered a group. As in figure 5a there are 5 groups. Visually, we can define that group 2 and group 3 belong to the lane marking of the middle lane. Distance between group 2 and group 3 is d, because distance between group 1 and group 2 closely equals with d, distance between group 3 and group 5 closely equals with d. So, group 1 and group 2 belong to the lane marking of the left lane, and group 3 and group 5 are on the lane marking of the right lane. Finally, the lane markings are shown as in figure 5b.

IOP Conf. Series: Materials Science and Engineering **323** (2018) 012026 doi:10.1088/1757-899X/323/1/012026

a. The lane marking estimation

b. Lane markings indicated by red line segments

Figure 5. The lane marking estimation.

3. Vehicle detection and tracking

3.1. Vehicle candidate detection

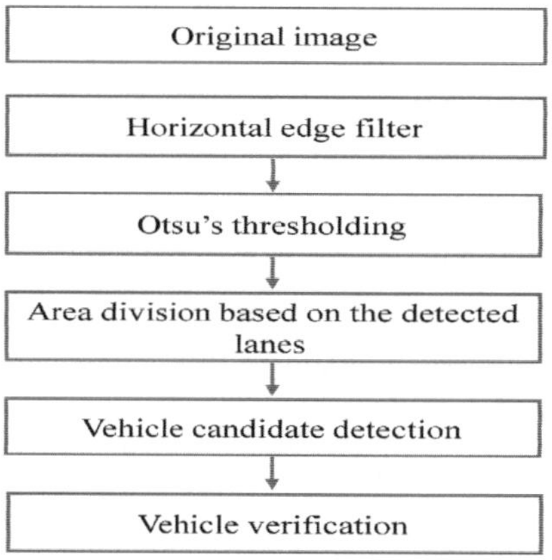

Figure 6. Algorithm for vehicle detection.

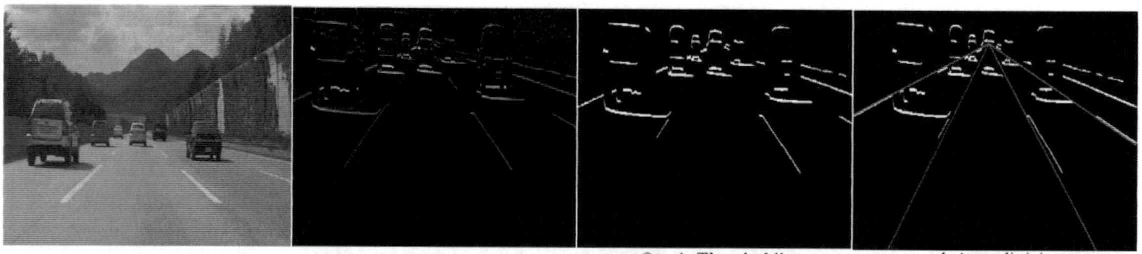

a. Original image b. Horizontal edge detection c. Otsu's Thresholding d. Area division

Figure 7. Demonstration for image processing algorithm.

Vehicle detection is the most important problem in this paper. Hence, the proposed method is shown in figure 6, and which is based on the detected lane markings from the previous section. In order to approach this feature, the original image in figure 7a is processed through horizontal edge filter shown in figure 7b. Then, we convert the received image from grayscale to binary as in figure 7c using otsu's thresholding. After having the binary image, we divide image into lane areas based on the detected lane markings. Information of the lane areas as in figure 7d is used to detect vehicle.

Figure 8a is the binary image of the left-side lane, Figure 8b presents the sum of the horizontal pixel following the vertical axis. From the graph in figure 8b, we can detect the candidate information.

Equivalently, candidates can be defined on the middle lane and right-side lane as shown in figure 8c and 8d.

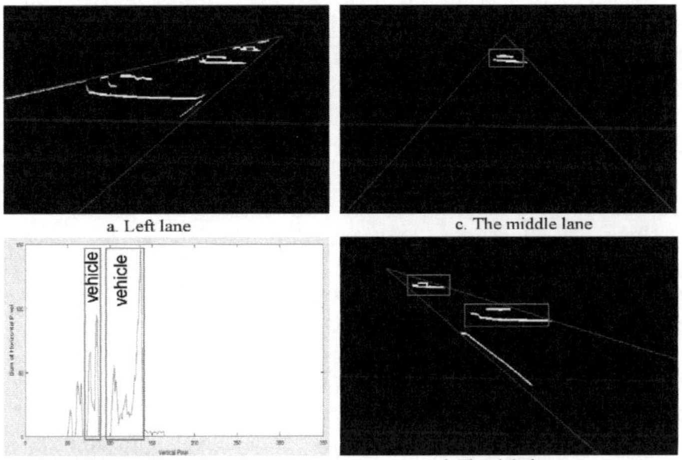

a. Left lane c. The middle lane

b. Vehicle candidate detection d. The right lane

Figure 8. Vehicle candidate estimation.

3.2. Vehicle verification

a. Original image 1 b. Horizontal edge 1 c. Original image 2 d. Horizontal edge 2

Figure 9. Noise on highway.

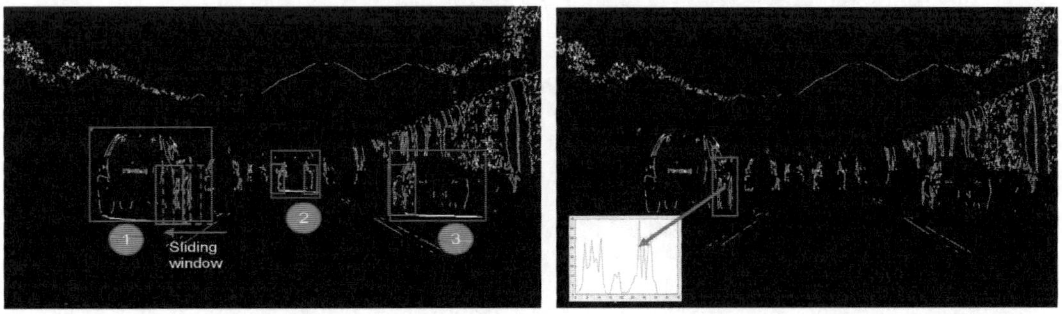

a. Sliding window for verification method b. Sum of pixel following the vertical axis

Figure 10. Verification method of candidates.

By using horizontal edge to detect vehicles, some wrong cases happened in test processing. In the detail, many noise appears from long highway such as shadow of bridges, shadow of the tension poles, or markings of repaired road in shown in figure 9. They affect our method.

In order to eliminate noise effect, because the back of vehicles has straight edge structure, we compare the detected candidates with this feature of vehicles. Figure 10a shows three objects which were identified from previous section, one of them is on the left lane, another is on the middle lane, and the left one is on the right lane. To verify them, the sliding window are created to check the vertical edge. For candidates on the left lane, the sliding window is used to check the vertical edge on the right side of candidates, for candidates on the right lane, the sliding window is used to check the vertical edge on the left side of candidates, and with objects on the middle lane, the vertical edge is checked on the both side of candidates as shown in figure 10a. If sum of the pixel of the sliding window following vertical axis is enough large as shown in figure 10b, we can conclude that these candidate are vehicle, on the contrary, they are not vehicle.

4. Experiment and Discussion

4.1. Lane detection

Before performing the experiments, we test the line detection algorithm and vehicle detection algorithm with various conditions such as rainy or foggy weather, and inside of a tunnel. The EDLines algorithm is used for testing in the three cases of conditions stated in figure 11. The line segments are detected clearly and do not affect by various conditions, line segments on the road markings are big and not separate. Therefore, we can use these segments to detect lanes by using the proposed method.

Figure 11. EDLines algorithm with the various conditions.

Figure 12. Result for lane detection.

Figure 12 shows some experiments for lane detection. Normal cases are represented in figure 12a in which lane is detected in good weather. In figure 12b, figure 12c, there are some noise which consists of cross-lines, white/yellow lines, direction indicating lines on the road, and objects which are on the road. Noise directly affects accuracy of the algorithm, making the system unstable. However, result of the proposed algorithm has turned out to be really good. Figure 12d is lane detection inside the tunnel.

4.2. Vehicle detection

In this paper, we show three scenes for vehicle detection. There are three vehicles in front of the host vehicle as shown in figure 13a and 13b. If the host vehicle wants to change lane to either left, or right side, it has to detect vehicles on the front and on the left lane or the right. The result of detection is shown in these figure. Figure 13c shows the detected vehicles on the road which has two lanes, and figure 13d shows the detected vehicle in tunnel.

Figure 13. Result for vehicle detection.

5. Conclusion

In this paper, we proposed the approach to detect lanes, detect the multi-vehicles for the driver assistance system .For lane detection, to detect lane in real-time, we use EDLines algorithm which can detect line segments between 10 ms and 20 ms. Our algorithm detects three lane areas of the frontal view. Based on these areas combined with horizontal edge feature of vehicles, vehicles are detected in each lane. From the outcome of experiments, our method obtains the goal in driver assistance system or warning system.

Acknowledgments

This work was supported by the Development Program for Economic Cooperative industry through the Ministry of Trade, Industry and Energy and the Korea Institute for the Advancement of Technology (R0003972).

This work was supported by the Human Resource Training Program for Regional Innovation and Creativity through the Ministry of Education and National Research Foundation of Korea (2015H1C1A1035919).

References

[1] http://www.who.int/mediacentre/factsheets/fs358/en/
[2] http://money.cnn.com/2016/01/20/luxury/volvo-no-death-crash-cars-2020/index.html
[3] G Dezhi, L Wei, D Jianmin and Z Banggui 2009 In. Conf. on Au. and Log. 980-85.
[4] C Rotaru, T Graf, and . Zhang 2008 J. of Real-Time I. Pro. 3, 311–22.
[5] J Heechul, M Junggon, and Kim Junmo 2013 IEEE In. Ve. Sym (IV).
[6] M He and X MEI 2012 In. Conf. on Wa. Ac. Me. Tech. and In. Pro.
[7] H Sungji, H Youngjoon and H Hernsoo 2009 W. a. of s. en. and tech 59, 455-59.
[8] M Atibi, I Atouf, M Boussaa and A Bennis 2015 In. Conf. on Ad. Wi. In. and Com. Tech 73, 24-31.
[9] N Mohamed, M Ahmad and S Swamy 2014 12th In. N. Cir. and Sys. Conf.
[10] P MinWoo, P Jungpil and J SoonKi 2013 Pro. of the 2013 ACM Re. in A. and Con. Sys, 171-76.
[11] C Akinlar and C Topal 2011 In. Conf. on I. Pro.

IOP Conf. Series: Materials Science and Engineering **323** (2018) 012027 doi:10.1088/1757-899X/323/1/012027

Interdigitated electrodes (IDE) using elastomer functionalized multi-walled carbon nanotube (MWNT) nanocomposites for the detection of oil spills

Tae-Yong Kim[1], Hyunseung Lee[1], Boyeong Park[1] and Yangsoo Kim[1,2,*]

[1]Department of Nanoscience & Engineering, Inje University, 50834 Gimhae, Korea
[2]High Safety Vehicle Core Technology Research Center, Inje University, 50834 Gimhae, Korea

*Email: cheykim@inje.ac.kr

Abstract. IDE using elastomer functionalized MWNT nanocomposites was fabricated and applied for the oil sensor. Two types of nanocomposites (i.e., MWNT-grafting-polyisoprene and Si-MWNT/natural rubber) having elastomer content of ~70 wt% were used as a sensing material of oil sensor and the relative electrical resistance response of the sensor to three different kinds of oils (i.e., gasoline, engine oil and pump oil) was investigated. Among three kinds of oils, the response of MWNT-grafting-polyisoprene nanocomposites IDE sensor to gasoline was the most significant. It showed the maximum relative electrical resistance of 18 within 20 min of dropping of 200 μL gasoline at room temperature for the MWNT-grafting-polyisoprene nanocomposites, which is far superior in sensitivity to the experimental results reported by Ponnamma et al in 2016 [1]. On the other hand, the response of Si-MWNT/natural rubber nanocomposites IDE sensor to gasoline was not appreciable. The elastomer functionalized MWNT nanocomposites prepared by "grafting-from" method, which is MWNT-grafting-polyisoprene in this study, is an excellent candidate material for the detection of oil spills.

1. Introduction

Accidents involving tankers, cargo ships, offshore platforms or drilling rigs and wells result in the release of significant amounts of crude oil and then it leads to the severe environmental disaster caused by soil and marine pollution. Oil spills can be controlled by expensive and time-consuming remedies such as chemical dispersion, combustion, mechanical containment and/or adsorption. However, monitoring oil spills with the help of sensing devices is the first step to ensure regulatory compliance. A lot of researches on oil sensors have been reported based on the measurements of UV fluorescence, IR reflection, ultrasonic, light scattering, electromagnetic absorption, conductivity, capacitance and other properties. These sensor devices have problems such as large size, large energy consumption, low sensitivity and high cost. Therefore, reliable and inexpensive oil sensor operating at room temperature needs to be developed.

There are noted three interesting research reports using electrically conductive elastomer nanocomposites of MWNT [1], polyaniline [2] and graphite [3] for the detection of oil spills. It has been known that carbon nanotube (CNT) based polymer nanocomposites are promising material components for the design of smart chemical sensor. Remarkable features of using CNTs in nanocomposite materials are the quite low percolation threshold (i.e., < 5 wt%) due to the high aspect ratio greater than 1000 and the appearance of excellent reinforcing effect for electrical, mechanical and thermal properties in nanocomposites. An essential prerequisite condition to maximize the material properties while minimizing the amount of CNTs added is the uniform dispersion of CNTs in the matrix without agglomeration caused by the van der Waals interaction forces between them. It has been known that polymer functionalization of CNT surface is an effective strategy for not only providing uniform dispersion but also suppressing the agglomeration of CNTs in polymer matrix. In this study, two different types of elastomer functionalized MWNT nanocomposites, which are MWNT-grafting-PI (polyisoprene) by "grafting-from" method and Si-MWNT/NR (natural rubber) by "grafting-to" method, were prepared and applied for oil sensor. It is the first experimental study on the IDE sensor using elastomer functionalized MWNT nanocomposites for the detection of oil spills, which has never been reported so far.

2. Experimental

2.1. Preparation of elastomer functionalized MWNT nanocomposites

Figure 1 shows the schematic procedures for the preparation of MWNT-grafting-PI by "grafting-from" method and Si-MWNT/NR by "grafting-to" method. The application of the activators regenerated by electron transfer for atom transfer radical polymerization (ARGET-ATRP) process [4] regarding the "grafting-from" method and the covalent bonding of NR to the silane treated MWNT [5] regarding the "grafting-to" method was made in this study.

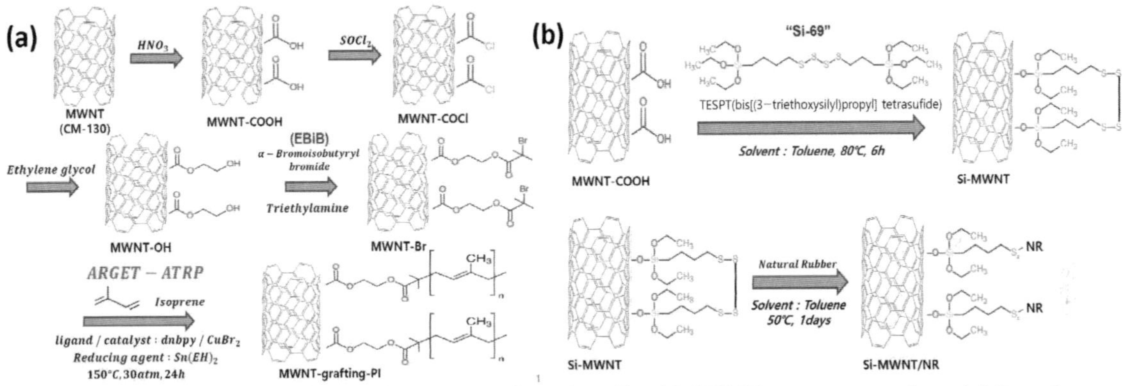

Figure 1. Preparation flow of elastomer functionalized MWNT nanocomposites: (a) "grating-from" method and (b) "grafting-to" method.

It was tried the preparation of PI functionalized MWNT nanocomposites (MWNT-grafting-PI) having a maximum content of PI in order to promote the sufficient swelling of PI when being contacted with oil. Characterization of nanocomposites was carried out by the application of thermogravimetric analyser (TGA) and vibrational viscometer, and also the C-Br content of MWNT-Br sample was estimated by the application of specific peak area analysis tool (e.g., "Origin-Polygon Area") regarding the corresponding X-ray photoelectron spectroscopy (XPS) spectrum.

2.2. Fabrication of IDE device and its application for oil sensor

A fabrication flow of IDE device and the working principle of oil sensor are schematically shown in figure 2. Pattern deposition on the glass substrate was carried by using indium as a metal source and the fabricated IDE pattern showed the electrode width of 300~350 μm and the electrode spacing of 250~270 μm. It was investigated the response of elastomer functionalized MWNT nanocomposites IDE sensor to three different kinds of oils, which are automotive fuel (i.e., gasoline), automotive engine oil (i.e., 5W30) and rotary pump oil (MR-100, Moresco). The electrical resistance measurements were performed with the electrometer (Keithley 6517B) in two electrode mode.

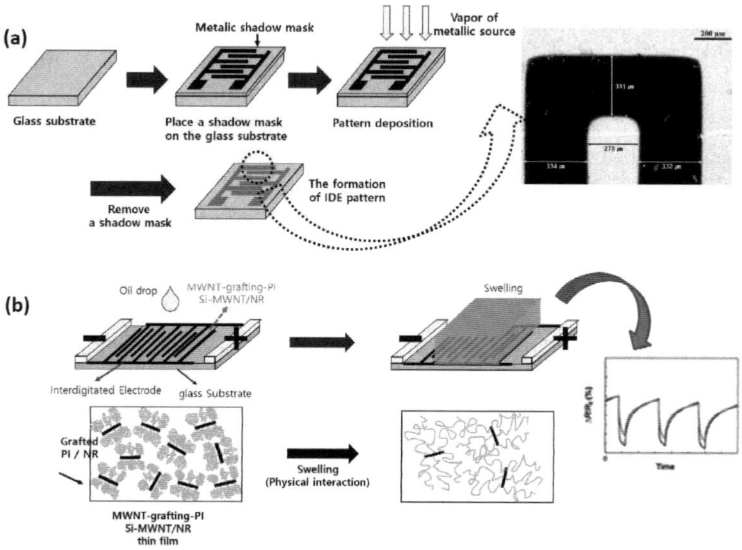

Figure 2. (a) Fabrication flow of IDE device and (b) working principle of IDE oil sensor using elastomer functionalized MWNT nanocomposites.

3. Results and discussion

Figure 3 shows how to control the elastomer (i.e., PI) content in MWNT-grafting-PI nanocomposites by using the combined analysis of TGA, viscometer and XPS.

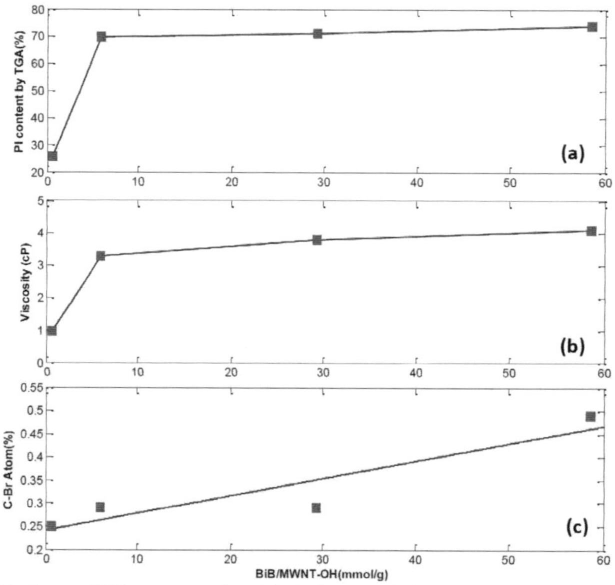

Figure 3. Variation of PI content by (a) TGA and (b) viscosity regarding MWNT-grafting-PI nanocomposites and also (c) C-Br Atom% of MWNT-Br depending on the BiB content based on the weight of MWNT-OH sample.

The MWNT-Br intermediate is used as an initiator and it reacts with isoprene, and then it leads to the preparation of MWNT-grafting-PI nanocomposites by the application of ARGET-ATRP process ("grafting-from" method). The MWNT-Br intermediate is prepared by the reaction of MWNT-OH intermediate with α-bromoisobutyryl bromide (BiB), and therefore the amount of C-Br bonds present in MWNT-Br can be controlled by the variation of the ratio of BiB to MWNT-OH, which is

[BiB]/[MWNT-OH] (mmol/g). The Br element covalently bonded to MWNT is used as a polymerization site of isoprene and thus it can be made a theoretical anticipation of the higher C-Br atom(%) value the larger polymerization yield. Figure 3-(c) shows a linear dependence of C-Br atom(%) value on the increase of [BiB]/[MWNT-OH]. However, both the PI content by TGA (figure 3-(a)) and the viscosity value by viscometer (figure 3-(b)) show a plateau state above 5.86 of [BiB]/[MWNT-OH]. In the experimental condition of this study, the PI functionalized MWNT nanocomposites (MWNT-grafting-PI) having more than 70 wt% of PI content were successfully prepared.

The preparation of NR functionalized MWNT nanocomposites (Si-MWNT/NR) was made by the application of "grafting-to" method and the NR content was adjusted to ~70 wt% using the TGA analysis based on the residual weight percent at a specified temperature. The quantitative estimation of elastomer content in the nanocomposites prepared in this study was made using TGA thermograms of MWNT-grafting-PI and Si-MWNT/NR shown in figure 4.

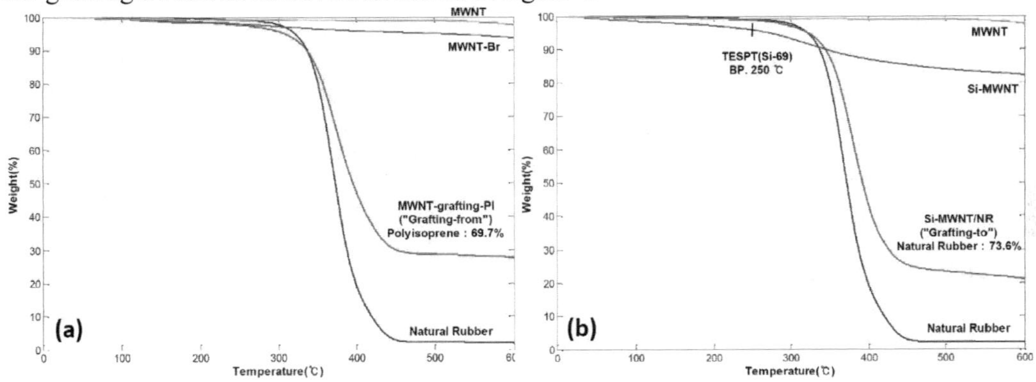

Figure 4. TGA thermograms of (a) MWNT-grafting-PI and (b) Si-MWNT/NR.

It was made the investigation of response of elastomer functionalized MWNT nanocomposites IDE sensor to the exposure of a specified amount of oil drop and the experimental results are shown in figure 5.

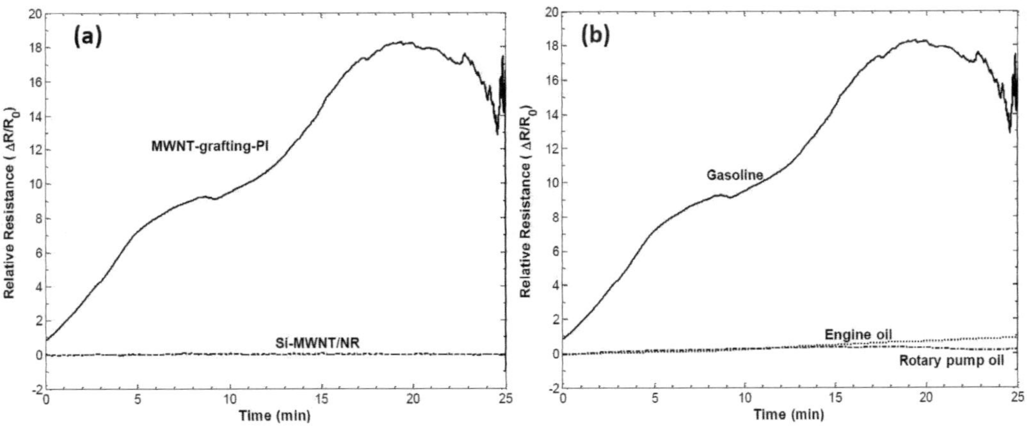

Figure 5. Response of (a) elastomer functionalized MWNT nanocomposites IDE sensor to gasoline and (b) that of MWNT-grafting-PI nanocomposites IDE sensor to three different kinds of oils.

Two types of composites showed quite different sensor responses in figure 5-(a). The relative electrical resistance, which is defined by the equation of $\Delta R(=R-R_0)/R_0$ in this study, for the PI functionalized MWNT nanocomposites (MWNT-grafting-PI) steadily increased with the oil exposure time and reached the maximum level of ~18 at 19 min of oil dropping of 200 μL gasoline. Ponnamma et al [1] reported the response of relative electrical resistance for the MWNT-filled NR composites prepared by a solution mixing of MWNT and chloroform to hydraulic oil, in which the maximum relative electrical resistance value was less than 1.3 at 15 min. The relative electrical resistance response for the NR functionalized MWNT nanocomposites (Si-MWNT/NR) was marginal showing less than 0.03. It reveals that the elastomer functionalized MWNT nanocomposites prepared by the "grafting-from" method provide far superior sensor sensitivity in detecting oil spills to the nanocomposites prepared by the "grafting-to" method, which could be ascribed to the relatively uniform dispersion and distribution of MWNTs in elastomer (polyisoprene).

Figure 5-(b) shows the relative electrical resistance response of MWNT-grafting-PI IDE sensor to three different kinds of oils. It is noted that the sensitivity to gasoline is the most noticeable among them and therefore the MWNT-grafting-PI IDE sensor represents the excellent sensor selectivity in detecting gasoline.

4. Conclusions

It is certainly feasible developing a reliable and inexpensive oil sensor operating at room temperature to prevent a catastrophic disaster caused by the oil spills involved in marine accidents. A practical strategy suggested in this study is the utilization of the elastomer functionalized MWNT nanocomposites as a sensing material for the detection of oil spills along with IDE device, and also the nanocomposites need to be prepared by "grafting-from" approach in order to obtain the uniform dispersion and distribution of MWNT in elastomer. How to maximize the PI content in PI functionalized MWNT nanocomposites (MWNT-grafting-PI) was proposed and the estimated content of PI was 69.7 wt% based on the TGA thermogram. The MWNT-grafting-PI nanocomposites IDE sensor showed not only the quite significant relative electrical resistance response to gasoline but also the outstanding sensor selectivity for gasoline among three different kinds of oils.

Acknowledgement
This work was supported by 2017 Daeheung Rubber & Technology Industry University Research Grant.

References

[1] D. Ponnamma, K. K. Sadasivuni, S. Thomas, I. Krupa, M. A. S. Al-Maadeed, Rubber Chem. Technol., 89, 306 (2016)
[2] K. K. Sadasivuni, D. Ponnamma, P. Kasak, I. Krupa, M. A. S. Al-Maadeed, Mater. Chem. Phys., 147, 1029 (2014)
[3] I. Krupa, M. Prostredný, Z. Špitalský, J. Krajči, M. A. S. Al-Maadeed, Smart Mater. Struct., 23, 125020 (2014)
[4] B. Wang, Z. Wang, F. Jiang, H. Fang, Z. Wang, RSC Adv., 4, 26468 (2014)
[5] Y. Nakaramontri, C. Kummerlöwe, C. Nakason, N. Vennemann, Polym. Comp., 36, 2113 (2015)

IOP Conf. Series: Materials Science and Engineering **323** (2018) 012028 doi:10.1088/1757-899X/323/1/012028

Study on the effect of sink moving trajectory on wireless sensor networks

Peijun Zhong [1,*] and Feng Ruan [2]

[1] ZTEsoft Technology Co. Ltd., Nanjing 210012, China
[2] School of Information & Control, Nanjing University of Information Science & Technology, Nanjing 210044, China

*Email: zhong.peijun@ztesoft.com

Abstract. Wireless sensor networks are developing very fast in recent years, due to their wide potential applications. However there exists the so-called hot spot problem, namely the nodes close to static sink node tend to die earlier than other nodes since they have heavier burden to forward. The introduction of mobile sink node can effectively alleviate this problem since sink node can move along certain trajectories, causing hot spot nodes more evenly distributed. In this paper, we make extensive experimental simulations for circular sensor network, with one mobile sink moving along different radius circumference. The whole network is divided into several clusters and there is one cluster head (CH) inside each cluster. The ordinary sensors communicate with CH and CHs construct a chain until the sink node. Simulation results show that the best network performance appears when sink moves along 0.25 R in terms of network lifetime.

1. Introduction

Recent advancement in micro-electronics, computing and wireless communication has enabled the fast development of low-power and multifunctional sensor nodes, which can integrate sensing, processing and communication functions in their tiny volumes. These tiny sensor nodes can be randomly deployed over an area of interest to work collaboratively. Besides, they can self-organize themselves based on their local collaboration to form wireless sensor networks (WSNs) [1].

Energy efficiency and energy balancing is one of the primary research challenges to the successful application of WSNs. Sensor nodes are usually powered with limited batteries and they cannot be easily recharged after deployment. Besides, hotspot problem can be caused by unbalanced energy consumption among sensors, especially when sink node is static. Under such circumstance, sensor nodes close to sink node will have more traffic to forward and they tend to deplete their energy much quickly than other sensors which are far away from sink [2].

By adopting sink mobility technology [3-4], the following advantages can be achieved. First, the hotspot problem can get alleviated as sink node moves around and energy consumption can get balanced among sensor nodes. Second, the whole network energy consumption can get reduced with relatively shorter transmission distance between sensor and sink. Third, the transmission latency can be shortened and network throughput can be increased with proper mobile sink design. Finally, sink mobility technology can ensure good network connectivity even under sparse networks.

In this paper, we mainly study the effect of sink moving trajectory on wireless sensor networks, with special focus on network lifetime (or remaining nodes alive). We divide the whole network into several clusters with same size. In each cluster there is a cluster head (CH) and ordinary sensor nodes will communicate with cluster head while CHs construct a chain until the mobile sink. Extensive simulations are performed to study the effect of sink moving trajectory on WSNs.

2. Related Work

LEACH(Low-Energy Adaptive Clustering Algorithm) [5] is a typical hierarchical routing protocol, where the sensor nodes are divided into two types, cluster heads (CHs) and ordinary sensor nodes. A sensor node delivers the sensor data to its related CH and CHs take charge of forwarding and fusing the sensor data to the base station. CHs are randomly selected and each node could only be chosen once in a certain amount of rounds. PEGASIS (Power-Efficient Gathering in Sensor Information Systems) [6] is an enhanced chain-based hierarchical protocol. In PEGASIS, each node transmits data package to the nearest neighbor which is closer to the base station than the source node. Several chains could be generated by this greedy algorithm, and the leader of the chain which is close to the base station takes the responsibility to forward the package to the base station.

A scheme called Velocity Energy-efficient and Link-aware Cluster-Tree(VELCT) is proposed [7]. The scheme contains two parts: set-up phase and steady-state phase. In set-up phase, the CHs are elected via threshold value and clusters form with better performance in an intra cluster communication. Then a few nodes are selected as Data Collection Node(DCN) to construct Data Collection Tree(DCT). In steady-state phase, CHs collect information from its members and forward the aggregated data to DCN and then the sink. Simulation results shows that it has a better performance in throughput, energy consumption and network latency than popular algorithms.

The authors proposed a scheduling mechanism based on spanning graphs by introducing mobile sink [8]. A heuristic path planning algorithm is presented to find a shortest path to avoid the obstacles. Mobile sink walks along the path and collect data from CHs via single-hop transmission. The results demonstrate that the algorithm extends the lifetime of the network effectively.

3. Our Proposed Routing Algorithm

3.1. Migration strategy of the mobile sink

In the algorithm, the direction and speed of the mobile sink are predetermined. The mobile sink moves around the center of the circle with a constant speed and a constant radius. Hence, the sink node only needs to broadcast its position in the initial parse. After certain time, the mobile sink moves to new position.

3.2. Clustering and CHs selection

Differ from other algorithms, we divided the whole network into n parts and every part is a cluster. Each sensor node calculates which cluster it belongs to via its position. At the beginning of each round, CHs selection will be conducted. The parameters of the selection are the residual energy of each nodev and the distance between one node and the mobile sink. The node close to the mobile sink broadcasts its weight in the cluster and claims a tentative CH. Only the nodes with higher weight broadcast its weight and become a new tentative CH. Finally, the node with the maximal weight is selected as the CH in each cluster.

3.3. Intracluster communication

There are two methods for a cluster member to communicate with its CH: direct transmission and multi-hop transmission. Any cluster member can communicate with its cluster head directly and the energy consumption is calculated using the following Formula 3:

$$E_1(S_i,CH_{S_i}) = \begin{cases} lE_{elec} + l\varepsilon_{fs} d(S_i, CH_{S_i})^2, d(S_i, CH_{S_i}) < d_0 \\ lE_{elec} + l\varepsilon_{mp} d(S_i, CH_{S_i})^4, d(S_i, CH_{S_i}) \ge d_0 \end{cases} \quad (1)$$

IOP Conf. Series: Materials Science and Engineering **323** (2018) 012028 doi:10.1088/1757-899X/323/1/012028

When cluster member has a long distance away from its cluster head, a relay node is chosen to forwarding the data package.

3.4. Formation of CH chain

In order to avoid long-distance communication between CHs and mobile sink, a chain of CHs is used. The process of the chain generation is divided into the following steps:

1) Mobile sink broadcast a message to require all the CHs to report their ID and position.
2) The CH which is closest to the mobile sink is chosen as the chain head.
3) Each CH transmits the fused date package to its neighbor CH using greedy algorithm.
4) Data transmission begins at the end of the chain and finally the chain head forwards the data to the mobile sink.

The result of chain generation is shown in figure 1.

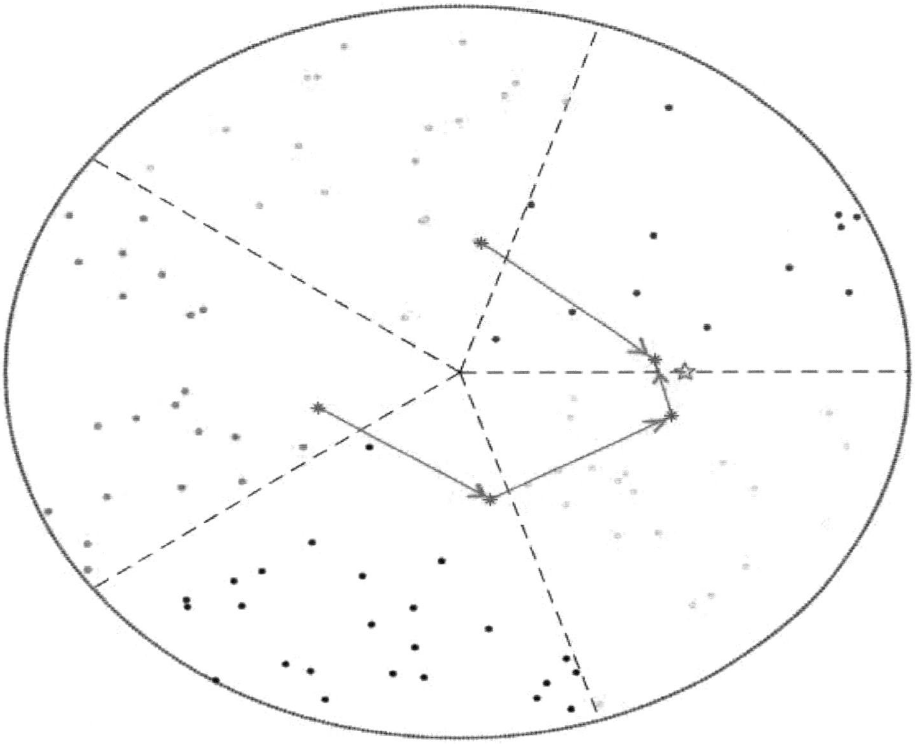

Figure 1. Chain generation.

4. Performance Evaluation

4.1. Simulation environment

In order to test our proposed algorithm, we use a MATLAB simulator to perform the simulation. The radius of the mobile sink is changed to explore the optimal performance. We also test our algorithm for WSNs with different network topologies from about 20 times.

4.2. Study on mobile sink radius

The trajectory of the mobile sink has a great influence on the performance of the network. We assume the mobile sink keep static(r=0) at the centre of the circle or moves around the center of the circle with the radius of 0.25R, 0.5R and R respectively. Simulation results are shown in figure 2.

From figure 2, we can clearly see that when the radius of the mobile sink equals about one fourth (0.25R) the network size, the network has the longest lifetime under different network topologies.

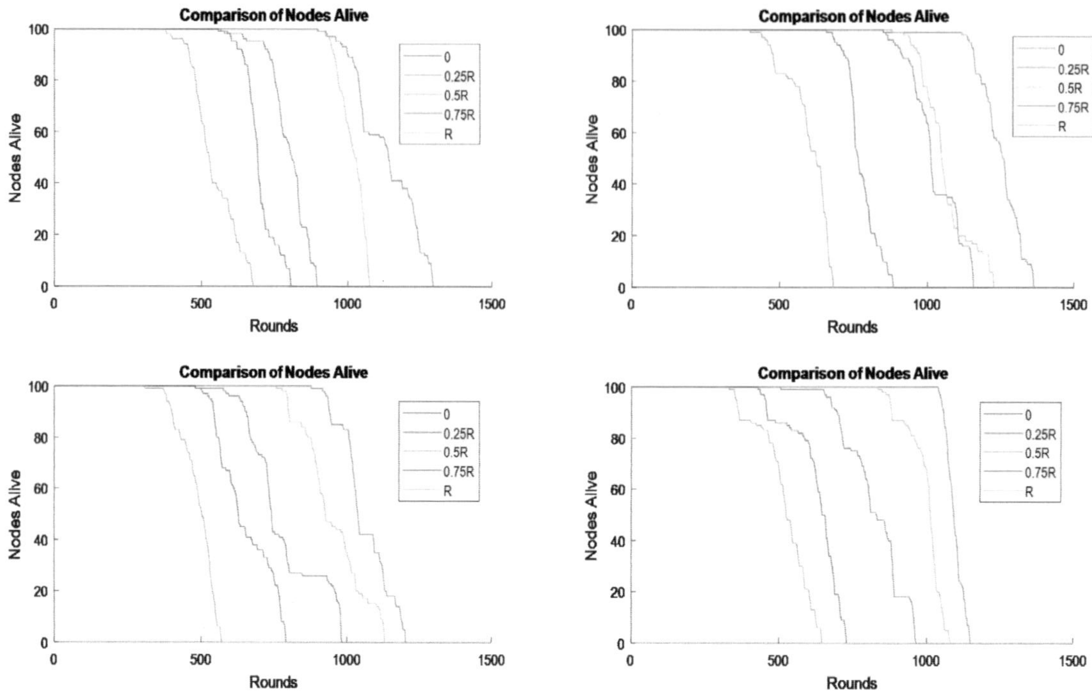

Figure 2. Different radius of mobile sink.

4.3. Study on different network size

We assume the network size is 150, 200, 300, 400 meters respectively. With the increase of the network size, the communication distance increases simultaneously. The lifetime of the network becomes short as the network size increases, while the mobile sink still has a better performance in terms of lifetime when its radius is one fourth R, as is shown in figure 3.

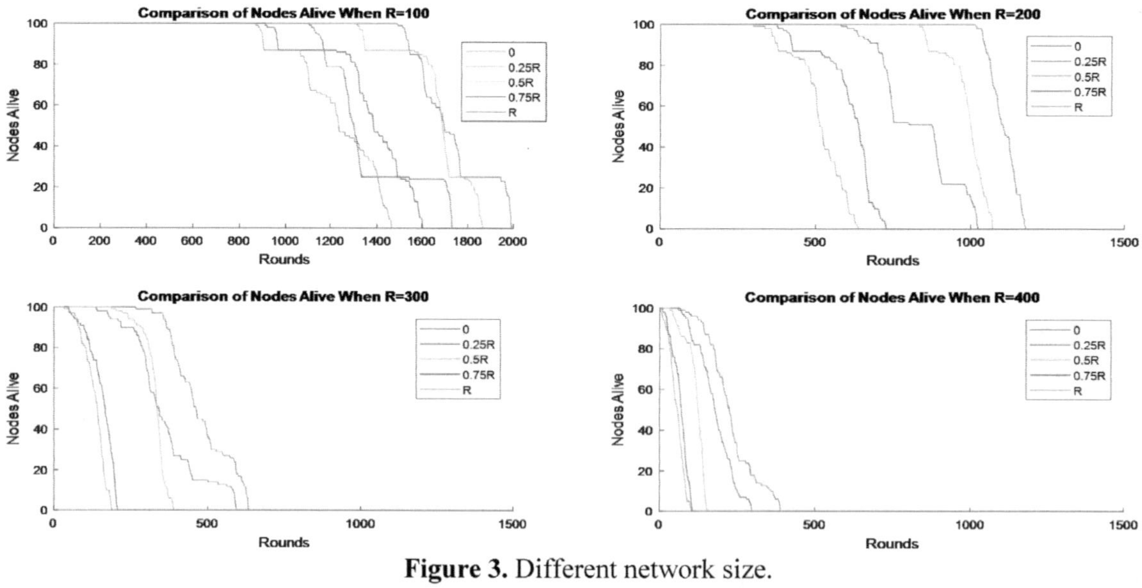

Figure 3. Different network size.

5. Conclusions

Mobile sink technology provides WSNs an alternative to alleviate the hot spot problem. In this paper, we combine the clustering method together with mobile sink strategy to improve network

IOP Conf. Series: Materials Science and Engineering **323** (2018) 012028 doi:10.1088/1757-899X/323/1/012028

performance. Based on our extensive simulations, we observe that when sink node moves along 1/4 radius of circular network, it has best performance in term of network lifetime. In the near future, we will extend our simulation by considering multiple mobile sinks as well as other network metrics such as energy consumption, latency etc.

References

[1] Akyildiz, I.F., Su, W., Sankarasubramaniam, Y., Cayirci, E. Wireless sensor networks: A survey, Computer Network, 2002, 38, 393–422.

[2] Jamal N. Al-Karaki, and Ahmed E. Kamal, Routing techniques in wireless sensor networks: a survey, IEEE Wireless Communications, December 2004, 11, 6-28.

[3] Luo J., Hubaux J. P., Joint Mobility and Routing for Lifetime Elongation in Wireless Sensor Networks, IEEE INFOCOM, 2005.

[4] Chatzigiannakis I., Kinalis A., and Nikoletseas S., Efficient Data Propagation Strategies in Wireless Sensor Networks Using a Single Mobile Sink, Computer Communications, 2008, 31(5): 896-914.

[5] Heinzelman, Rabiner W, Chandrakasan, et al., Energy-efficient communication protocol for wireless microsensor networks. Ad hoc & Sensor Wireless Networks, 2000, 18(10) pp. vol.**2**.

[6] Lindsey S., and Raghavendra C.S., PEGASIS: power efficient gathering in sensor information systems. Proc.Int'l Conf. Comm. (ICC '01), 2001.

[7] Velmani R, Kaarthick B. An Efficient Cluster-Tree Based Data Collection Scheme for Large Mobile Wireless Sensor Networks. IEEE Sensors Journal, 2015, 15(4): 2377-2390.

[8] Xie G., Pan F., Cluster-Based Routing for the Mobile Sink in Wireless Sensor Networks with Obstacles, IEEE Access, 2016, 4:2019-2028.

IOP Conf. Series: Materials Science and Engineering **323** (2018) 012029 doi:10.1088/1757-899X/323/1/012029

An energy efficient multiple mobile sinks based routing algorithm for wireless sensor networks

Peijun Zhong [1,*] and Feng Ruan [2]

[1] ZTEsoft Technology Co. Ltd., Nanjing 210012, China
[2] School of Information & Control, Nanjing University of Information Science & Technology, Nanjing 210044, China

*Email: zhong.peijun@ztesoft.com

Abstract. With the fast development of wireless sensor networks (WSNs), more and more energy efficient routing algorithms have been proposed. However, one of the research challenges is how to alleviate the hot spot problem since nodes close to static sink (or base station) tend to die earlier than other sensors. The introduction of mobile sink node can effectively alleviate this problem since sink node can move along certain trajectories, causing hot spot nodes more evenly distributed. In this paper, we mainly study the energy efficient routing method with multiple mobile sinks support. We divide the whole network into several clusters and study the influence of mobile sink number on network lifetime. Simulation results show that the best network performance appears when mobile sink number is about 3 under our simulation environment.

1. Introduction

With the fast development and progress of micro-electro-mechanism system (MEMS), system on chip and wireless communication technologies, a large number of tiny, and low cost sensors are produces. These nodes are randomly placed in a certain area, collecting the data of surroundings, transferring them to the sink nodes by single hop or multiple hops, and forming wireless sensor networks (WSNs). In recent years, WSNs are widely used in natural disaster monitoring, military tracking, agricultural observation, etc. [1].

Energy saving and balancing is always a hot and challenging research issue for WSNs [2]. Research shows that sensor nodes near the static sink nodes will have heavier transmission burden than other ordinary node, which make themselves hot spots and this phenomenon is known as energy holes problems [3].

In order to better utilize the energy of each node in the wireless sensor network and to balance their energy consumption as much as possible, the concept of mobile sink nodes is proposed [4-5]. Those sink nodes can move at a certain speed in the network at a specified or random path, and collect data from the nearby nodes during the mobile process. In this way, because the sink nodes are moving constantly, there will not be some nodes being always close to the sink nodes to become hot spots and the energy consumption of nodes around sink nodes is also balanced.

2. Related Work

An energy efficient routing protocol based on clustering method was proposed [6], which was applicable to WSNs with obstacles. The mobile sink moved from the starting point to the cluster head, collecting data from the cluster head through single hop transmission, and completed a round of data collection to return to the starting point. Experimental results indicated that the network lifetime was effectively extended, and the scheduling mechanism solved the complexity of scheduling problems in networks with obstacles efficiently. The authors proposed a low energy consumption routing protocol for the constrained problem of sink nodes in sensor networks [7]. It transferred the data to the nearest next arrival location of the mobile sink node, which ensured that the data transmission path is shortest. This protocol had high robustness and low energy consumption and is suitable to be used in networks of delayed tolerance.

The authors devised an optimization framework for mobile data collection where mobile data collection was carried out by two steps [8]. First, a part of sensor nodes were selected to transmit data directly to the mobile sink node. Second, the distributed algorithm was designed to adjust the data rate and link scheduling to improve the network utilization. The authors presented a new method for maximizing the throughput of data collection [9]. First, data collection was carried out using mobile collector with fixed mobility and an optimization model was proposed, which included effective time and heterogeneous time of sensor transmission in each time slot. Second, the maximum amount of time slot available for each sensor node was determined. Last, they confirmed the constant speed of the mobile collection node and designed an online centralized algorithm to deal with the NP problem.

Another method of increasing network throughput called an opportunity data collection method was proposed [10]. The authors first put forward a new routing metric named Contact-Aware ETX, and evaluated the transmission delay caused by the packet retransmissions. Then the routing standards and routing protocols were defined to realize the CA-ETX. Finally, by combining CA-ETX with dynamic routing, an opportunity backpressure collection was proposed. The tests stated clearly that the methods proposed in the above two papers could effectively improve network throughput and reduce network latency.

3. System Model

3.1. Network Model

In this paper, we deploy N sensor nodes in the circular field the radius of which is R, denoted as: {N1,N2…….,Nn} separately The sensor field is divided into several sectors evenly and each sensor node belongs to a sector based on its position.

3.2. Energy Model

We adopt the first radio energy model as the energy model to calculate energy consumption. Energy consumption contains two parts: transmission consumption and reception consumption.

Energy consumption for transmission l-bit message is shown in formula 1:

$$E_{Tx}(l,d)=\begin{cases} l\cdot E_{elec}+l\cdot \varepsilon_{fs}\cdot d^2, & if d<d_0 \\ l\cdot E_{elec}+l\cdot \varepsilon_{mp}\cdot d^4, & if d\geq d_0 \end{cases} \quad (1)$$

where E_{elec} denotes the energy consumption to run transmitter or receiver circuit. ε_{fs} and ε_{mp} denotes the amplification coefficient for the free space model and the multi path fading model.

Reception consumption can be calculated as is shown in formula 2:

$$E_{Rx}(l)=l\cdot E_{elec} \quad (2)$$

4. Our Proposed Routing Algorithm

4.1. Clustering phase

We divided the whole network into n parts and every part is a cluster. Each sensor node calculates which cluster it belongs to via its position. At the beginning of each round, CHs selection will be conducted. The parameters of the selection are the residual energy of each node and the distance between node i and mobile sink. The node close to the mobile sink broadcasts its weight in the cluster and claims a tentative CH. Only the nodes with higher weight broadcast its weight and become a new tentative CH. Finally, the node with the maximal weight is selected as the CH in each cluster.

4.2. Intra-cluster communication phase

Usually, a cluster member will communicate with its CH in two ways: direct transmission and multi-hop transmission. Any cluster member can communicate with its cluster head directly and the energy consumption is calculated using the following Formula 3:

$$E_1(S_i,CH_{S_i}) = \begin{cases} lE_{elec} + l\varepsilon_{fs} d(S_i, CH_{S_i})^2, d(S_i, CH_{S_i}) < d_0 \\ lE_{elec} + l\varepsilon_{mp} d(S_i, CH_{S_i})^4, d(S_i, CH_{S_i}) \geq d_0 \end{cases} \quad (3)$$

When cluster member has a long distance away from its cluster head, a relay node is chosen to forwarding the data package.

5. Performance Evaluation

5.1. Simulation Environment

In order to test our proposed algorithm, we use a MATLAB simulation environment. The radius of the mobile sink is changed to explore the optimal performance. We also test our algorithm with different WSNs network topologies for about 20 times.

5.2. Study on Mobile Sink number

We research the effect of mobile sink number to the lifetime of the network via simulations. We only use single mobile node in the beginning and then increase the number of mobile sink once a time. Simulation results show that with the increase of the mobile sink number, the performance of the network in terms of lifetime improves, as is shown in figure 1. We also attentions that the trend of the lifetime slows down with the mobile sink increased. As the mobile sinks are much more expensive than ordinal sensor nodes, 3 mobile sinks are most suitable.

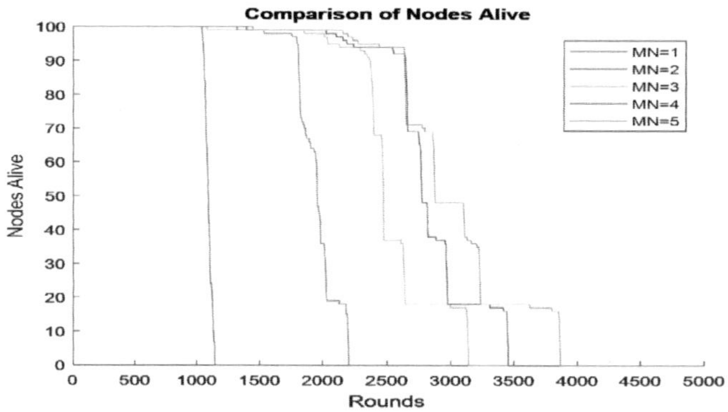

Figure 1. Different mobile sink number.

5.3. Study on different methods of weight calculation

It is a common method to select cluster heads through the residual energy of nodes, but the nodes chosen by this way are usually far away from mobile sink which have high residual energy. This will increase the number of long distance communication between CHs and mobile sink, increasing the energy consumption of the entire network. Therefore, in order to achieve a balance between residual energy and the distance from CHs to mobile sink to prolong the lifetime of the system. We calculate the weight of each node via using residual energy dividing distance between node and the mobile sink as a standard for choosing a cluster head, Simulation results are shown in figure 2, and the latter approach has a better performance.

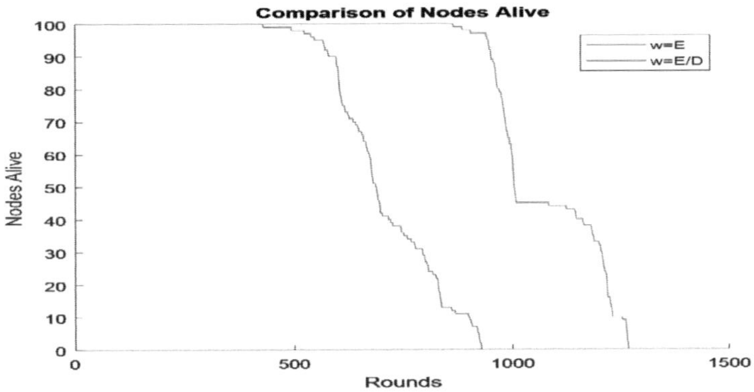

Figure 2. Different ways to calculate weight.

6. Conclusions

During the design of energy efficient routing algorithms for WSNs, how to alleviate the hot spot problem is an important research issue. In this paper, we proposed an energy efficient clustering algorithm for WSNs, with mobile sink support. We first give the system model and our hierarchical routing algorithm. Then, we present simulation results to validate the performance of our proposed algorithm. Simulation results show that network has the best performance when mobile sink number is around 3. Even though we continuously increase cluster number, there is no big increase while the cost from sink node will increase which is not necessary. Also, we evaluate the selection criteria of two metrics during the selection of CHs. In the near future, we will make further study on the mobile sink parameters such as moving velocity, other clustering methods on WSNs performance.

References

[1] I. F. Akyildiz, W. Su, Y. Sankarasubramaniam, E. Cayirci, A Survey on Sensor Networks, IEEE Communications Magazine, 2002, 40(8):102-14.

[2] G. Anastasi, M. Conti, M. Di Francesco, and A. Passarella, Energy conservation in wireless sensor networks: A survey, Ad Hoc Netw., 2009, 7(3): 537–68, 2009.

[3] S. Olariu and I. Stojmenovic, Design Guidelines for Maximizing Lifetime and Avoiding Energy Holes in Sensor Networks with Uniform Distribution and Uniform Reporting, IEEE INFOCOM, 2006.

[4] Y. Gu, F.J. Ren, Y. Ji, and J. Li, The Evolution of Sink Mobility Management in Wireless Sensor Networks: A Survey, IEEE Communications Surveys and Tutorials, 2016, 18(1):507-24.

[5] J. Luo, J. P. Hubaux, Joint Mobility and Routing for Lifetime Elongation in Wireless Sensor Networks, IEEE INFOCOM, 2005.

[6] G. Xie, F. Pan, Cluster-Based Routing for the Mobile Sink in Wireless Sensor Networks With Obstacles. IEEE Access, 2016, 4: 2019-28.

[7] M. T. Nuruzzaman, Ferng H W. A low energy consumption routing protocol for mobile sensor networks with a path-constrained mobile sink, 2016 IEEE International Conference on Communications (ICC'16). 2016.

[8] C. Wang, S. Guo, Y. Yang, An Optimization Framework for Mobile Data Collection in Energy-Harvesting Wireless Sensor Networks, IEEE Transactions on Mobile Computing, 2016, 15(12): 2969-86.

[9] A. Mehrabi, K. Kim, Maximizing Data Collection Throughput on a Path in Energy Harvesting Sensor Networks Using a Mobile Sink, IEEE Transactions on Mobile Computing, 2016, 15(3): 690-704.

[10] S. Yang, U. Adeel, Y. Tahir, J.A.McCann, Practical Opportunistic Data Collection in Wireless Sensor Networks with Mobile Sinks, IEEE Transactions on Mobile Computing, 2016, 16(5): 1420-33.